Cured of cancer

from childhood to adulthood
quality of survival

Omslagontwerp en ontwerp binnenwerk: René Staelenberg, Amsterdam

ISBN 90 5356 666 X
NUR 870

Printed and bound by CPI Group (UK) Ltd, Croydon, CR0 4YY

Cured of cancer

from childhood to adulthood
quality of survival

Academisch proefschrift

Ter verkrijging van de graad van doctor
aan de Universiteit van Amsterdam op gezag van de
Rector Magnificus prof. mr. P.F. van der Heijden
ten overstaan van een door het college voor promoties ingestelde
commissie, in het openbaar te verdedigen in de Aula der Universiteit
op donderdag 15 mei 2003, te 12.00 uur

door

Neeltje Elisabeth Langeveld

geboren te Den Helder

Amsterdam University Press

Promotiecommissie:

Promotoren: Prof. dr. P.A. Voûte
                 Prof. dr. R.J. de Haan

Co-promotor: Dr. M.A. Grootenhuis

Overige leden: Prof. dr. H.S.A. Heymans
                 Prof. dr. F.E. van Leeuwen
                 Dr. M.A.G. Sprangers
                 Prof. dr. G.A.M. van den Bos
                 Prof. dr. T. van Achterberg

Faculteit Geneeskunde

Contents

Chapter 1

Introduction

Introduction

Over the past decades, paediatric oncologists, paediatric oncology nurses and other health-care professionals caring for children with cancer have witnessed profound changes in the nature of their practice. Whereas care of the child with cancer was 30 or 35 years ago a matter of controlling pain and keeping the child as comfortable as possible, with little expectation regarding long-term cure, many of those diagnosed now have an excellent chance of long-term, disease-free survival.

As paediatric oncology was evolving, and the word 'cure' was entering the physicians and nurse's vocabulary, previously unexpected late effects of childhood cancer and its treatment began to be identified [1]. There was growing recognition that both chemotherapy and radiation therapy could have adverse effects upon normal body tissue that may manifest themselves months or even years after completion of treatment. They included every body system and varied in severity from relatively benign conditions such as radiation-induced alopecia to potentially life-threatening problems such as radiation-related breast carcinoma or anthracycline-induced cardiomyopathy [2].
Possible harmful effects on the long-term mental health of these patients came also into focus. The patients and their families have gone through a severely traumatic period, even under the best of circumstances. Living under a life threat has altered life style, relationships and future plans and has caused more subtle intrapsychic change as well [3].

Both the potentially life-saving treatment of a child with cancer and the emerging reports on late effects raised serious concerns about the health status and quality of life (QL) of the survivors of childhood cancer. What about the adult such a child would become? Will a person who experienced a cancer during childhood become a "normal" adult and would he or she be able to make his or her way in society? Would he or she be capable of leading a full life, without handicap, and with intellect and sexual functions intact? These questions, asked by health-care professionals in the late 1970's, remain so today, even though there is a growing amount of literature documenting the late psychosocial effects of childhood cancer and its treatment. Despite all research efforts, the research to date on psychosocial outcomes of survivors of childhood cancer has been characterised by equivocal and, at times contradictory findings [4].

The quality of survival of young adult survivors of childhood cancer is the main subject of this thesis. To put the quality of survival of these young persons in perspective, it is important to understand the medical aspects of the disease, including the major late effects of cancer in children and the impact of the treatment period on the child. In this introductory chapter a general overview is given of the most important epidemiological aspects of childhood cancer. The somatic late effects of both the disease and the therapy are described, together with the impact of the treatment period on the child. Furthermore the need for follow-up and the concept of QL are discussed. Finally, an outline of this thesis is given.

Epidemiological aspects of childhood cancer

Incidence
Paediatric and adolescent cancer cases only represent a small proportion of the world-wide cancer burden [5]. Within western industrialised countries, the proportion of cancers occurring within the paediatric age range is approximately 2%. While cancer in children is rare, it is the second leading cause of death in children and the primary cause of death from disease [6]. The characteristics of malignant tumours in childhood differ greatly from those in adults, not only in the types and histology, but also in the anatomic locations of tumours. The most common cancers of adults, such as lung, female breast, stomach, large bowel, prostate and skin, are extremely rare among children. Children develop malignancies of rapidly growing body systems, such as the reticuloendothelial system, the central nervous system, and connective tissue. Internationally, there are striking differences in the incidences of all cancers for children between the ages of 0 and 14 years [6]. Possible explanations for this difference are among others the quality of the medical care system and its ability to diagnose cancer, differences in the classification of various neoplasm's and the thoroughness of the cancer surveillance and reporting system.

In the Netherlands approximately 400 children to the age of 16 years are diagnosed with cancer every year [7]. About a third of all childhood cancers are leukaemias, predominantly acute lymphoblastic leukaemia (ALL). Brain and spinal tumours are the second most diagnostic group. Lymphomas and non-Hodgkin's lymphomas have a somewhat higher incidence than Hodgkin's disease. Neuroblastoma and Wilms' tumour, the two most frequent embryonal tumours of childhood, each account for 6-7% of registrations, as do soft-tissue sarcomas, while retinoblastoma accounts for 3%. Nearly all of the remaining cases are bone tumours, germ-cell tumours, and epithelial tumours, like malignant melanoma, skin carcinoma and thyroid carcinoma.

The age-incidence distribution varies between diagnostic groups [8]. There is a peak in incidence in the younger age groups for many of the childhood cancers (i.e. leukaemias and tumours of the central nervous system, neuroblastoma, retinoblastoma, Wilms' tumor, and hepatoblastoma) and the increasing incidence with age in others (i.e. Hodgkin's disease, bone tumours, thyroid cancer, and melanoma) [6]. Diagnoses of childhood cancer are not equally present in boys and girls, for example leukaemias and lymphomas occur more often in boys.

Aetiology
In spite of a great deal of research, there is still little known about the cause of most types of cancer in childhood. Aetiology has been connected to various environmental carcinogenetic agents such as radiation, drugs, chemicals, and viruses, the strongest causal evidence being shown for prenatal exposure to radiation [9]. Certain chromosal anomalies and other inherited conditions are also associated with an increased risk of developing cancer in early life, and this supports the hypothesis that a genetic predisposition to cancer may underline many such cancers. The association of acute leukaemia with Downs' syndrome is perhaps the best known. However, although there may be an underlying genetic predisposition in many children, this does not imply that childhood cancer is likely to recur in subsequent children in a family.

Survival rates
Since the 1960's, major advancements have been made in the treatment and cure of childhood cancer [6]. Before this time only a few children could be cured by surgery, sometimes with the addition of radiation therapy. It was not until the introduction of chemotherapy that the situation really began to improve. Centralisation of care, improved supportive care, and large successful clinical trials have led further to the improved survival rates in paediatric oncology. We have now reached the stage where between 60 and 70 percent of all children diagnosed with cancer can be cured. For some types of cancer the prognosis is extremely good, for example Wilms' tumour and Hodgkin's disease have cure rates of up to 90 percent. However, there are other cancers, for example, advanced neuroblastoma, where the outlook is still poor.

Treatment
The major modalities of cancer therapy - surgery, radiation therapy, and chemotherapy- are the same for paediatric and adult malignancies. Usually different treatment modalities are combined. The treatment for any child depends on a number of factors, such as the type of cancer, its location, the degree of growth and spread of the tumour, and other prognostic factors. Most children are treated in prospective clinical trials, called treatment protocols, which are closely co-ordinated and monitored. Receiving therapy on a protocol is of major benefit for any patient because it assures that the patient will receive the latest peer-reviewed therapies. A long tradition of clinical protocols has prevailed in paediatric oncology. Early in the fight against childhood cancer, co-operative groups were formed, which employ uniform protocols for their patients, increasing the total of children studied and allowing comparisons of how patients respond to one or another treatment program. In Europe, there is one major paediatric co-operative group, the International Society of Paediatric Oncology (SIOP), initiated in 1971. Their Wilms' Tumour Clinical Trial and Study successfully demonstrated that a randomised clinical trial could be used to correlate clinical grouping of a tumour with treatment options and survival. The second and following SIOP Wilms' tumour trials formulated questions based on information generated by the first study. Similar groups like SIOP have emerged world-wide, like Children's Oncology Group (COG), United Kingdom Childhood Cancer Study Group

(UKCCSG), and Societé Francaise d'Oncologie Pediatrique (SFOP). The widespread practice of protocol participation among paediatric oncology patients is an important reason why there has been remarkable progress made in curing children with cancer.

Unfortunately, the treatment of cancer in growing children is not without side effects and children with cancer will experience many of these unavoidable effects during treatment. Side effects are often referred to as early or late. Side effects that occur within days to weeks of treatment are called early. Side effects that occur months to years later are called late side effects.

Early side effects
Early side effects can occur after any type of treatment, including surgery, chemotherapy and radiation therapy. The severity of these side effects depends on the type of cancer being treated, the location of the disease, the age of the child, and the intensity of the treatment.

Exploratory surgery may be performed for several paediatric cancers at diagnosis, or after therapy to determine tumour response. The use of surgery increases the risk of intestinal obstruction and the development of adhesions. Surgical removal of the spleen, especially at a young age, results in an increased risk of serious life-threatening blood infection with Haemophilus influenza or Streptococcus pneumoniae. In the past, bone tumours involving an arm or a leg were commonly treated with amputation. Although amputation controlled the local spread of tumour, the procedure resulted in functional and cosmetic problems that often seriously affected the child's quality of life. Advances in diagnostic imaging and in the use of chemotherapy before surgical therapy have now permitted the more frequent use of limb-sparing procedures in children with bone tumours.

Chemotherapy is generally non-specific for tumours, in that the effect of the drugs is not confined to tumour cells. In fact, these drugs have an effect on all rapidly dividing cells, including normal cells, which accounts for many of their side effects, such as hair loss, nausea and vomiting, mouth sores, diarrhoea, fever and bone marrow depression resulting in anaemia, leukopaenia, and thrombocytopenia. Certain chemotherapy agents have specific toxic effects on certain parts of the body. For example, cyclophosphamide and ifosfamide can cause bladder damage with bleeding, whereas mercaptopurine can result in liver damage.

Radiation destroys tumour tissue, but it also can damage nearby normal tissue. Short-term effects can include tissue damage similar to a burn, skin discoloration, or weakness. All side effects generally relate to the area or region treated. For example, treatment of the pelvis may cause diarrhoea, but it cannot cause alopecia. Conversely, treatment of the head may cause alopecia, but it does not cause diarrhoea.

Late side effects
The frequency, severity, nature, and timing of the development of somatic late effects depend on many factors, including the location and size of the primary

tumour, extent of the tumour, intensity of the local therapy, type of therapy, and physiologic status of the child [10]. Late effects may manifest themselves in different ways, including 1) clinically obvious effects that interfere with activities of daily living (e.g. pulmonary fibrosis resulting in respiratory distress); 2) clinically subtle effects noticeable to the trained observer (e.g. learning impairment after treatment of the central nervous system); and 3) subclinical effects detectable only by laboratory screening or x-ray studies (e.g. elevated liver enzyme levels) [11].

Surgery can have long-term effects, particularly if it is disfiguring. For example, the impact of surgical procedures such as enucleation, amputation, and hip disarticulation is lifelong. The major long-term effects from chemotherapy and radiation therapy will be discussed in the following paragraph.

Second cancers
The development of a second malignant neoplasm (SMN) is probably the most feared consequence of therapy. Overall, for children who have experienced a primary malignancy, the incidence of new neoplasm's ranges between 8% and 12% at 20 years. The general risk of SMNs is approximately 10 to 15 times greater than the risk of first neoplasm's in the general population [12]. Factors that determine potential risk include certain chemotherapeutic agents, radiation therapy, genetic predisposition, and other genetic conditions such as neurofibromatosis [2].

Hypothalamic-Pituitary glands
Cranial and facial radiation therapy can result in deficiencies of hormones produced by the brain, such as the pituitary and hypothalamic hormones. These hormones include growth hormone and stimulating hormones that control the function of the thyroid, ovaries, testes, and adrenal gland. Hypothalamic-pituitary injury may result in a reduced growth rate, obesity, and/or early, delayed, or arrested onset of puberty [2].

Thyroid dysfunction
Hypothyroidism is the most common abnormality reported after radiation therapy to the neck. Other complications observed after thyroid irradiation include the development of thyroid nodules, hyperthyroidism and thyroid cancer [2].

The gonads
In boys, testicular function is very sensitive to the effects of both radiation and chemotherapy [13]. Among chemotherapy drugs, alkylating agents are associated with infertility, the risk of which is correlated with dose. In girls, radiotherapy and chemotherapy can also alter ovarian function. Cases of premature menopause are reported [14].

Cardiac dysfunction
Acute and/or chronic cardiovascular sequelae occur in children treated with mediastinal radiation, anthracyclines, and/or cyclophosphamide [2,15]. Mediasti-

nal radiation has at least three separate effects on the heart. First, thickening of the atrioventricular valves and pericardium may cause mitral or tricuspid insufficiency or restrictive pericardial changes. Second, histologic changes in the coronary arteries occur and predispose these children to early coronary atherosclerosis and early myocardial ischemia and infarction [16]. Third, there is often damage to the right ventricle (which is anterior to the chest and therefore receives the highest radiation dose) with areas of myocardial fibrosis [2].

Pulmonary toxicity
Chemotherapy, especially with bleomycin, busulfan and nitrosoureas (CCNU/BCNU), can cause pulmonary fibrosis. Concurrent lung irradiation increases the damaging effect [17]. Radiation therapy can cause hypoplasia, acute pneumonitis and chronic fibrosis [18].

Kidney, bladder and urinary tract toxicity
The kidney and bladder are both affected by radiation and chemotherapy. Chronic nephritis and cystitis are the most commonly noted effects. Chemotherapy, including dactinomycin, cisplatin, methotrexate, anthracyclines, and nitrosoureas, can cause renal failure or enhance the radiation effects on the urinary tract. Cyclophosphamide and ifosfamide causes haemorrhagic cystitis, bladder fibrosis, atypical bladder epithelium, and renal tubular dysfunction [2].

Gastrointestinal and liver damage
Radiation may enhance the risk of obstruction from adhesions in children who have undergone laparotomy and at high dose result in bowel stricture and retroperitoneal fibrosis [14]. Chronic radiation damage to the liver is recognised and could result in fibrosis and portal hypertension [19]. Hepatic fibrosis may also arise from long-term administration of methotrexate.

Musculoskeletal and soft tissue damage
Bones, soft tissue, and blood vessels are most vulnerable to radiation during periods of rapid growth. Therefore the effects on growth are most pronounced in children less than 6 years of age and during the pubertal growth [17]. Growth impairment may result in reduced or uneven growth, leading to scoliosis, short stature, and extremity deformities such as functional limitations and shortening of the extremity. Radiation of soft tissues also causes cosmetic deformities, like fibrosis and hypoplasia. The developing breast tissue may be damaged by relatively small doses of radiotherapy and this can result in hypoplasia and failure of lactation [14].

Haematological and immunological late effects
Persistent subclinical bone marrow damage can occur after conventional chemotherapy and similar changes are reported after radiotherapy [20]. Splenectomy is a long-term risk factor for sepsis and splenic damage from radiation therapy has been reported. Prolonged immunosuppression is a major issue for survivors of bone marrow transplantation, especially those with chronic Graft versus Host Disease.

Ears and eyes
Late radiation damage may affect all parts of the eye, the orbit and surrounding soft tissue. The most important consequences are cataract and dry eye. Cisplatin is a well recognised cause of hearing loss and the effect may be potentiated by radiation. There is an association between steroid therapy and cataract formation.

Teeth and salivary glands
Dental abnormalities occur after high-dose radiation to the head and neck region and include poor root development, incomplete calcification, delayed or arrested tooth development, and multiple caries [21]. Radiation directly to the salivary gland results in decreased secretions, causing mouth dryness.

Neurological and neuropsychological sequelae
Neuropsychologic deficits and neuroanatomic abnormalities can occur as a result of whole brain radiation and intrathecal (IT) chemotherapy. Neuropsychologic, or cognitive, impairments are typically manifested as significant declines in IQ and academic achievements scores and as specific deficits in visual motor integration, memory, attention and motor skills [22.23]. Non-verbal skills, like abstract reasoning, visual spatial skills and arithmetic, are particularly vulnerable to the damaging effects of radiation therapy and IT chemotherapy. Atrophy and decreased subcortical white matter are the most common neuroanatomic abnormalities after central nervous system treatment [2].

Psychosocial effects
In *Chapter 4*, an overview of the research on different aspects of QL (physical functioning, psychological functioning, social functioning and sexual functioning) in young adult survivors of childhood cancer is given.

The impact of the treatment period on the child

The diagnosis of cancer has an impact on everything in the lives of the children. They have to deal with a series of crisis: awaiting diagnosis; receiving the diagnosis of a life-threatening illness; beginning and progressing through treatment; and living with all of this uncertainty on a daily basis [24]. A variety of emotional reactions to a confirmed diagnosis of cancer such as shock, denial, grief, anger, and depression can be expected [25]. Experience suggests that the child and family require approximately a year to adapt to the change in life-style that may result from the diagnosis of cancer [26]. In the meantime, the children are often faced with repeated invasive medical procedures as part of their treatment or as a method for evaluating the effectiveness of their treatment. In the case of leukaemia, for example, they must undergo repeated lumbar punctures to administer chemotherapy and bone marrow aspirations to determine the status of the disease. The treatment period is a period primarily characterised by multiple hospitalisations and frequent clinic visits, which are for most children anxiety-producing. The child must not only cope with a life-threatening illness, but he must also experience separation from family life, friends and school, and side effects from medications and other treatment that often seem more stressful than the disease itself.

How children react to this crisis situation may be influenced by pre-illness emotional strengths and weakness developed as individuals and as a family unit [25]. The child's developmental level of understanding for example, will influence its ability to comprehend what is taking place. Some factors have been identified that may affect how well a child adjusts to having cancer. Younger children tend to have a better and more rapid adjustment back to a normal life than their adolescent counterparts. Having a shorter treatment course, without relapse, also tends to produce a better psychological outcome for the child. Other factors affect all the family members' ability to cope with a child's cancer. Everyone will cope better if there is a high level of family cohesiveness and support, when there are few stresses in other areas of life (finances, work, or other family problems) and when there is open communication that allows the family to discuss a wide range of issues [24].

The need for follow-up

Looking at the impact of the disease, its treatment, and the consequences from that, it is evident that cancer in childhood leaves an inedible imprint on all whose lives are touched by it, especially the child. Together with the high possibility of long-term survival or even cure, it is not strange that the goals of treatment now include considerations for the quality of life and follow-up of the patients post-treatment. In 1996, SIOP developed guidelines for the care of childhood cancer survivors and stressed the importance of psychological support and the education of patients regarding a healthy lifestyle. They stated: "We advocate the establishment of a specialty clinic oriented to the preventive medical and psychosocial care of long-term survivors......... The goal is to promote long-term physical, psychosocial, and socio-economic health and productivity, not merely to maintain an absence of disease or dysfunction" [27]. As a result of these and other recommendations and the growing awareness of the needs of survivors of childhood cancer, some institutions began late-effects clinics using a multidisciplinary team to monitor and support survivors. These follow-up clinics not only provide comprehensive care for the survivors, but also participate in research projects that may improve the quality of life for current and future long-term survivors.

In February 1996, The Emma Kinderziekenhuis/Academic Medical Center (EKZ/AMC) in Amsterdam began a long-term follow-up clinic to monitor long-term sequelae of childhood cancer and its treatment. The EKZ/AMC is one of the 5 major paediatric oncology centers in the Netherlands, where children with cancer are treated. An average of 200 children (from 0-18 years) with cancer is referred annually to this hospital for consultation, diagnostic examination and treatment. Since approximately 1968, the EKZ/AMC has offered facilities for the systematic care of children with cancer. Indefinite surveillance of survivors has also been a long-standing policy. Since 1996, a model for providing comprehensive care to the survivors has been developed and a special outpatient clinic was set up for all former childhood cancer patients treated in the hospital. Patients become eligible for transfer from active-treatment clinics to the follow-up clinic when they complete cancer treatment succesfully at least 5 years earlier. Survivors are evaluated annually in the clinic by a paediatric oncologist (persons aged <18 years) or internist-oncologist (persons aged >18 years) for late medical effects, as well as a research nurse or psychologist for psychosocial effects.

The studies that are discussed in this thesis, took place at the follow-up clinic in the EKZ/AMC. With an increasing amount of children entering adult life who are cured of cancer during childhood, it is estimated that approximately one in 750-800 young adults in the Netherlands will be a survivor of childhood cancer [28]. Both the relevance and importance for current and future survivors as the potential implications for society of having such large numbers of survivors, decided us to evaluate the outcome of young adult survivors of childhood cancer. In particular, different aspects of QL were investigated.

The concept of Quality of Life

Assessment of QL is complicated by the fact that there is no universally accepted definition for QL. In the past, many researchers measured only one dimension, such as physical function, economic concern, or sexual function. More recently, researchers have attempted further definition of QL. The World Health Organization defines QL as "individuals' perception of their position in life in the context of the culture and value system in which they live and in relation to their goals, standards, and concerns"[29]. The definition includes six broad domains: physical health, levels of independence, psychological state, social relationships, environmental features, and spiritual concerns. The importance of this definition to childhood cancer survivors lies in the inclusion of both emotional and social dimensions of health in addition to physical aspects. While many survivors have no physical evidence of disease and appear to have made full recoveries, others have to come to terms with the chronic, debilitating, or delayed effects of therapy. All remain at risk of the development of late sequelae of the former disease and/or treatment and of second malignancies. Furthermore, in most cases the life-threatening experience of cancer is never forgotten. In many ways, survival enhances appreciation for life, while at the same time reminding survivors of their vulnerability. The metaphor of the Damocles syndrome illustrates this dichotomy and the way individual survivors interpret this metaphor for life will influence the quality of their survival [30].

Objectives and structure of this thesis

The main aim of this thesis was to study the quality of survival in a large cohort of Dutch long-term survivors of childhood cancer. The study population consisted of survivors who were more than 5 years after treatment without recurrence of malignant disease and were older than 16 years at the time of investigation.

The thesis is organised in the following order.
The study in *Chapter 2* reports on the medical and psychosocial problems encountered temporarily or permanently in the population of childhood cancer survivors who were evaluated in the long-term follow-up clinic. How these data

should guide the future development of risk-adapted follow-up programs will be discussed in some depth.

Hereafter the focus is changed to the quality of survival. In *Chapter 3*, an overview of the results of studies into the QL of young adult survivors of childhood cancer is presented. The concept of QL is defined, and, based on the literature, limitations of the studies and methodological difficulties are described. Finally, suggestions for future research are given.

In *Chapter 4*, the results of a qualitative study of fatigue in a small group of childhood cancer survivors are described. The purpose of this study was to explore the concept of fatigue from a survivors' perspective. We wanted to get more information about off-treatment fatigue in these survivors, because a better understanding of off-treatment fatigue is essential to help patients with fatigue and to develop intervention strategies to ameliorate these symptoms.

In *Chapter 5*, we investigated the issue of fatigue further and assessed the level of fatigue in a large cohort of survivors and compared the results with a group of young adults with no history of cancer. In addition, the impact of demographic, medical and treatment characteristics and depressive symptoms on survivors' fatigue is identified.

In *Chapter 6*, self-esteem, worries and the generic quality of life in survivors is investigated and put into perspective of that observed in a reference group. The contribution of various factors and self-esteem in explaining survivors' degree of worries and quality of life is identified and discussed.

In *Chapter 7*, we focus merely on the social aspects of quality of life and therefore educational achievement, employment status, living situation, marital status and offspring in survivors and a reference group is studied. Further, the influence of various factors associated with poor social functioning are identified.

Additionally, we explored the level of posttraumatic stress symptoms in our sample of survivors and the impact of demographic, medical and treatment factors on survivors' posttraumatic stress level. The results are described *in Chapter 8*.

Finally, in *Chapter 9*, the findings are summarised and a general discussion is presented. Future directions for research and clinical practice are formulated. An English and Dutch summary concludes this thesis.

References

1. Ruccione K. The role of nurses in late effects evaluations. In: Nesbit ME, editor. Late effects in successfully treated children with cancer. London, Philadelphia, Toronto: W.B. Saunders Company 1985, 205-221.
2. Hobbie W, Ruccione K, Moore IK, Truesdell S. Late effects in long-term survivors. In: Foley GV, Fochtman D, Mooney KH, eds. Nursing Care of the Child with Cancer. Orlando, Florida: W.B. Saunders Company 1993, 466-496.
3. Lansky SB, List MA, Ritter-Sterr C, Klopovich P, Chang P-N. Late effects: psychosocial. In: Nesbit ME, editor. Late effects in successfully treated children with cancer. London, Philadelphia, Toronto: W.B. Saunders Company 1985, 239-246.
4. Kazak AE. Implications of survival: Pediatric Oncology Patients and their Families. In: Bearison A, Mulhern R, eds. Pediatric Psychooncology. New York: Oxford University Press 1994: 171-192.
5. Muir C, Waterhouse J, Mack T. Cancer Incidence in five continents. IARC scientific publication no. 88. 1987. Lyon, International Agency for Research on Cancer Scientific Publications.
6. Smith MA, Gloeckler Ries LA. Childhood cancer: incidence, survival, and mortality. In: Pizzo PA, Poplack D, eds. Principles and Practice of Pediatric Oncology. Philadelphia: Lippincott Williams & Wilkins 2002: 1-12.
7. Paulides J, Kamps WA, Caron H. Childhood Cancer in the Netherlands 1989-1997. The Netherlands Cancer Registry 2000.
8. Stiller CA, Draper GJ. The epidemiology of cancer in children. In: Voute PA, Kalifa C, Barrett A, eds. Cancer in Children. Clinical Management. Oxford University Press 1998, 1-20.
9. Plon SE, Petersen LE. Childhood cancer, heredity, and environment. In: Pizzo PA, Poplack DG, eds. Principles and Practice of Pediatric Oncology. Philadelphia: Lippincott 1997: 11-36.
10. Late adversities of treatment in long-term survivors of childhood cancer. New York: American Cancer Society 1978.
11. Fochtman D, Fergusson J, Ford N, Pryor A. The treatment of cancer in children. In: Fochtman D, Foley GV, eds. Nursing care of the child with cancer. Boston: Little Brown and Company 1982, 177-231.
12. Meadows AT, Fenton JG. Follow-up care of patients at risk for the development of second malignant neoplasms. In: Schwartz CL, Hobbie WL, Constine LS, Ruccione KS, eds. Survivors of Childhood Cancer. Assessment and Management. St. Louis, Missouri: Mosby-Year Book, Inc. 1994, 319-328.
13. Mustieles C, Munoz A, Alonso M, Ros P, Yturriaga R, Maldonado S, Otheo E, Barrio R. Male gonadal function after chemotherapy in survivors of childhood malignancy. *Med Pediatr Oncol* 1995, 24, 347-351.
14. Hawkins MM, Stevens MC. The long-term survivors. *Br Med Bull* 1996, 52, 898-923.
15. Kremer L. Anthracycline cardiotoxicity in childhood cancer. Thesis. University of Amsterdam, the Netherlands 2001.
16. Hicks GL. Coronary artery operations in radiation-associated atherosclerosis: Long-term follow-up. *Ann Thoracic Surg* 1992, 53, 670-674.
17. Schwartz CL, Hobbie WL, Constine LS. Survivors of Childhood Cancer: Assessment and Management. St Louis, Missouri: Mosby-Year Book, Inc 1994.
18. Hudson MM. Late effects of cancer therapy. In: Steen G, Mirro J, eds. Childhood Cancer. A handbook from St.Jude Children's Research Hospital. Cambridge, Massachusetts: Perseus Publishing 2000: 491-503.
19. Barnard JA, Marshall GS, Neblett WW, Gray G, Grishan FK. Noncirrhotic portal fibrosis after Wilms tumour therapy. *Gastroenterology* 1986, 90, 1054-1056.
20. Mauch P, Constine L, Greenberger J, Knospe W, Sullivan J, Liesveld JL, Deeg HJ. Hematopoietic stem cell compartment: acute and late effects of radiation therapy and chemotherapy. *Int J Radiat Oncol Biol Phys* 1995, 31, 1319-1339.
21. Jaffe N, Toth B, Hoar RE, Ried HL, Sullivan MP, McNeese MD. Dental and maxillofacial abnormalities in long-term survivors of childhood cancer: Effects of treatment with chemotherapy and radiation to the head and neck. *Pediatrics* 1984, 73, 816-823.
22. Mulhern RK, Crisco JJ, Kun LE. Neuropsychological sequelae of childhood brain tumors: A review. *J Child Clin Psychol* 1983, 12, 66-73.

23. Moore IM, Kramer JH, Wara W, Halberg F, Ablin AR. Cognitive function in children with leukemia: Effect of radiation dose and time since irradiation. *Cancer* 1991, 68, 1913-1917.
24. Wiard S, Jogal S. The psychosocial impact of cancer. In: Steen G, Mirro J, eds. Childhood Cancer. A Handbook from St.Jude Children's Research Hospital. Cambridge, Massachusetts: Perseus Publishing 2000: 461-469.
25. Varni JW, Katz ER. Psychological aspects of cancer in children. A review of the research. *J Psychosoc Oncol* 1988, 5, 93-119.
26. Hall M, Havelin K, Conatser C. The challenges of psychological care. In: Fochtman D, Foley GV, eds. Nursing care of the child with cancer. Boston: Little Brown and Company 1982, 317-353.
27. Masera G, Chesler M, Jankovic M, Eden T, Nesbit ME, van Dongen-Melman J, Epelman C, Ben Arush MW, Schuler D, Mulhern R. SIOP Working Committee on Psychosocial Issues on Pediatric Oncology: Guidelines for Care of Long-term Survivors. *Med Pediatr Oncol* 1996, 27, 1-2.
28. Heikens J. Childhood cancer and the price of cure. Studies on late effects of childhood cancer. Thesis. University of Amsterdam, the Netherlands 2000.
29. World Health Organization DoMH. WHO-QOL Study protocol: The development of the World Health Organization quality of life assessment instrument. Geneva, Switzerland 1993.
30. Leigh SA, Stovall EL. Cancer Survivorship. Quality of Life. In: King CR, Hinds PS, eds. Quality of Life. From Nursing and Patient Perspectives. Sudbury, Massachusetts: Jones and Bartlett Publishers 1998, 287-300.

Chapter 2

Follow-up of long-term survivors of childhood cancer: 6 years experience with a specialised care and screening programme

Abstract

Purpose: To describe medical and psychosocial problems encountered temporarily or permanently in a group of 976 childhood cancer survivors seen at the long-term follow-up clinic of the Emma Kinderziekenhuis/Academic Medical Center.

Patients and methods: All childhood cancer survivors were seen at the outpatient clinic. Screening for late effects was performed according to protocols based on the previously used treatment modalities. At each visit a detailed medical history was taken and a complete physical examination was performed to allow for the identification of problems as previously unknown late treatment effects. Therefore, all data obtained from the screening procedures as well as detailed oncological treatment information were registered in an especially developed database.

Results: A total of 4004 medical or psychosocial problems were registered for the 976 survivors. The median number of problems was 3. Almost a quarter of the survivors, however, had maximally one problem registered. Those survivors who have survived a brain tumour or Ewing's sarcoma had the highest number of registered problems, survivors of Hodgkin's disease and acute lymphoblastic leukaemia the lowest.

Conclusion: Although it is clear that the majority of the childhood cancer survivors suffer from medical and/or psychosocial problems, additional risk estimates must be made on the complete cohort of survivors previously treated at our centre. These will lead to the development of risk-adapted follow-up programs. The usefulness of these risk-adapted programs must be communicated both to the survivors and those responsible for the health-care budgets. As primary treatment strategies continue to change, the logistics of our late effects clinic will allow for future evidence-based changes in the follow-up programs.

Introduction

Cancer in children is relatively rare [1]. Over the last decades survival rates have increased significantly [2,3]. In the Netherlands this has led to a currently estimated frequency of 1 out of 750-800 young adults being a childhood cancer survivor [4]. Since more than two decades it is recognised that both single modality and intensive combined modality treatments can lead to late adverse effects [5]. Knowledge about the epidemiology, pathophysiology and natural history of these late treatment effects has increased gradually [6-8]. The long term side effects nowadays seen in childhood cancer survivors include an increased mortality and cancer risk [9-17], infertility [8,18], heart failure [19-23], growth and bone mass abnormalities [24-27], endocrine disturbances [5,8,28], fatigue [29], as well as impaired cognitive functions and psychosocial problems [30-37].

The need for long-term follow-up of these childhood cancer survivors is uniformly recognised [38]. The form in which this follow-up should take place, however, is still under debate [39]. Furthermore, the possibilities to develop follow-up programs both now and in the future will depend on all kind of socio-economic, demographic and medical variables [39,40], and the wishes and the knowledge of the childhood cancer survivors [41,42].

In our hospital, the long-term follow-up clinic was established in 1996 as a co-operative effort between the Departments of Paediatric Oncology and Medical Oncology. The primary aim of our clinic is to provide adequate patient care. Secondary objectives are the development of a logistic system for additional research and the establishment of an adequate system for data capturing also allowing for further research. The aim of this study is to report the medical and psychosocial problems encountered in the first 976 childhood cancer survivors seen at our follow-up clinic. How these data should guide the future development of risk-adapted follow-up programs will be discussed in some depth.

Patients and methods

Since 1966 a registry (Werkgroep Kindertumoren (WKT) registry) is maintained in the Emma Kinderziekenhuis of children treated for malignancies, benign tumours and related diseases necessitating aggressive treatment modalities (e.g., Langerhans'cell histiocytosis). Patients become eligible for transfer from active-treatment clinics to the long-term follow-up clinic of the Emma Kinderziekenhuis/Academic Medical Center (EKZ/AMC) when they have completed cancer treatment successfully at least 5 years earlier. The follow-up clinic started in 1996 after a period of extensive preparations with the development of both the follow-up protocols and the data capturing system. Following the establishment of the follow-up clinic survivors were invited to attend after direct referral from the paediatric oncology outpatient clinic or after tracing previously discharged survivors by a variety of methods: 1) through medical records in our hospital, 2) by contacting general practitioners or other medical specialists, and 3) through municipal records. The first years mainly childhood cancer survivors over the age of 18 years were screened according to the protocols. The paediatric age group has been included regularly since spring 2000.

Survivors are screened annually or biennially at the clinic by a paediatric oncologist (persons aged <18 years) or medical oncologist (persons aged >18 years) for late medical effects. During the survivors' visit a detailed medical history is taken and a complete physical examination is performed. Additional investigations are ordered by protocols based on previously applied treatment modalities. For psychosocial effects they are seen at least once by a research nurse or psychologist. In case of encountered medical and/or psychosocial problems survivors are referred to other health care providers (e.g., general practitioners, medical specialists or psychologists) for further evaluation and treatment in the vicinity of the survivors' home.

Relevant data obtained from the screening procedures are registered in an especially developed database, PLEKsys. The PLEKsys database also contains

Table 1. Demographic and medical characteristics of the study group

Variable	Survivors (n=976)	
Sex	N	%
Men	519	53
Women	457	47
Age at follow-up (years)		
Mean/median ± SD	25 / 25 ± 7.1	
Age at diagnosis (years)		
Mean/median ± SD	7 / 7 ± 4.8	
Follow-up after start of treatment (years)		
Mean/median ± SD	18 / 18 ± 6.7	
Diagnosis	N	%
Acute lymphoblastic leukaemia	198	20
Acute non-lymphoblastic leukaemia	19	2
non-Hodgkin's lymphoma	123	13
Hodgkin's Disease	87	9
Rhabdomyosarcoma and other soft tissue sarcomas	92	9
Nephroblastoma	141	14
(ganglio)-neuroblastoma	48	5
Osteosarcoma	50	5
Ewing's sarcoma	33	3
All brain tumours	70	7
other	115	12

SD, standard deviation

Table 2. Medical and psychosocial problems diagnosed in survivors of different childhood cancers

diagnostic group	number of survivors	problems		
		mean	SD	median
Acute lymphoblastic leukaemia	198	2.66	2.99	2
Acute non-lymphoblastic leukaemia	19	3.47	3.03	3
non-Hodgkin's lymphoma	123	4.09	3.23	4
Hodgkin's Disease	87	3.29	3.25	2
Rhabdomyosarcoma and other soft tissue sarcomas	92	3.75	2.58	3
nephroblastoma	141	4.94	3.24	4
(ganglio)-neuroblastoma	48	5.21	3.78	4
osteosarcoma	50	3.98	2.44	3
Ewing's sarcoma	33	5.30	2.36	5
all brain tumors	70	6.96	4.18	6

SD, standard deviation

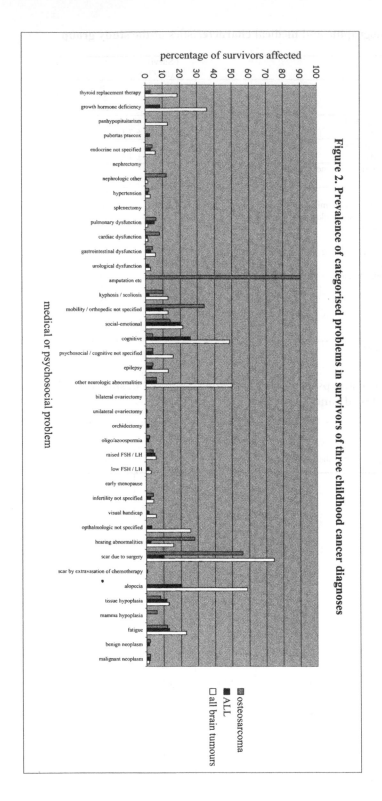

Figure 2. Prevalence of categorised problems in survivors of three childhood cancer diagnoses

Table 3. Categorisation (based on reference [6]) of grouped medical and psychosocial problems as registered in the PLEKsys database for the 976 childhood cancer survivors

Functional group	Problem	% of survivors affected	Functional group	Problem	% of survivors affected
endocrine	thyroid replacement therapy	5.8	infertility	bilateral ovariectomy	0.2
	growth hormone deficiency	5.7		unilateral ovariectomy	1.9
	panhypopituitarism	1.6		orchidectomy	2.9
	precocious puberty	0.9		oligo / azoospermia	4.7
	not specified	4.2		raised FSH / LH	10.8
organ toxicity	Nephrectomy	13.8		low FSH / LH	1.1
	nephrologic other	4.7		early menopause	0.4
	Hypertension	3.2		not specified	7.6
	Splenectomy	0.6	sensory	visual handicap	3.1
	pulmonary dysfunction	7.1		ophthalmologic ns	6.0
	cardiac dysfunction	4.5		hearing abnormalities	6.5
	gastrointestinal dysfunction	6.1	cosmetic	scar due to operation	54.8
	urological dysfunction	3.9		scar due to extravasation of chemotherapy	0.5
mobility / orthopedic	amputation etc	6.5		alopecia	11.9
	kyphosis / scoliosis	10.1		tissue hypoplasia	20.7
	not specified	13.9		mamma hypoplasia	1.9
psychological / cognitive	social-emotional	13.3	fatigue	fatigue	12.5
	Cognitive	14.0	subsequent neoplasm	benign	2.4
	not specified	4.7		malignant	2.6
neurologic	Epilepsy	3.3			
	not specified	12.2			

detailed oncological treatment information. For this study the database was queried for demographic characteristics (sex, age at diagnosis, time of follow-up, age at follow-up), diagnoses and encountered medical and or psychosocial problems. The problems were registered according to a predetermined list (Appendix 1). Problems were scored when: 1) a physical condition could unequivocally be observed, e.g., a scar resulting from an operation or an amputation, 2) a physical condition could be unequivocally determined from the patients' notes, e.g., a nephrectomy or splenectomy, and 3) a problem was present according to current medical and/or psychosocial standards. All problems experienced temporarily or permanently by the childhood cancer survivors were included. Descriptive statistics were performed using Microsoft Excel 2000 software.

Results

As off March 2002 nine hundred and seventy-six childhood cancer survivors have been seen and screened at our follow-up clinic. Survivor characteristics are shown in Table 1. One fifth of the survivors were previously treated for acute lymphoblastic leukaemia (ALL), and approximately equal numbers for non-Hodgkin's lymphoma (NHL) and nephroblastoma. All survivors were at presentation treated according to the institutional, national or international protocols that were relevant at that time. For the 976 childhood cancer survivors a total number of 4004 medical or psychosocial problems was registered. The median number of problems found per survivor was 3. There is an obviously skewed distribution with some survivors suffering from a large number of problems and almost a quarter of the survivors having maximally one problem registered (Figure 1).
Comparison of the number of problems found per primary diagnosis, showed that those previously treated for acute leukaemia and Hodgkin's Disease (HD) have the lowest average number of problems registered. Survivors of brain tumours and Ewing's sarcoma have the highest numbers of registered problems (Table 2).

In order to allow for comparison with other studies the problems were categorised according to a previously published system of functional groups [6]. The results of this categorisation are shown in Table 3. The problems most frequently observed (>10% of the patients) in the childhood cancer survivors were an operation scar, tissue hypoplasia, cognitive and social-emotional problems, fatigue, nephrectomy, elevated follicle stimulating hormone (FSH) and/or luteinising hormone (LH) and kyphosis/scoliosis. It was to be expected that not all primary cancer groups would encounter the same type of problems in the years following their cancer treatment. Figure 2 illustrates this for the diagnoses osteosarcoma, ALL and the group of the brain tumours. The brain tumour survivors more often suffer from cognitive and other neurological sequelae than those surviving ALL and osteosarcoma. In ALL survivors, however, the proportion of survivors suffering from a cognitive problem was still 26%. Besides the cognitive problems alopecia was frequently observed in both leukaemia and brain tumour survivors. Fatigue was seen in 12%, 13% and 23% of the survivors of osteosarcoma, ALL and brain tumours, respectively.

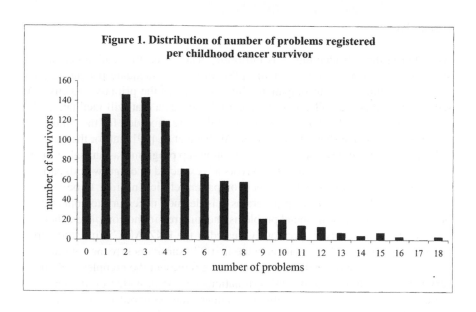

Figure 1. Distribution of number of problems registered per childhood cancer survivor

Discussion

As stated in the introduction, it is uniformly recognised that childhood cancer survivors should be followed for life [38]. Providing adequate patient care is, from the beginning of our outpatient clinic, viewed as the primary objective. A secondary goal is the collection of patient relevant data that will facilitate additional research aimed at the improvement of the care provided for the childhood cancer survivors. We share the opinion of Wallace et al. [39] that for the future it will be necessary to develop risk-adapted follow-up programs, as annual or biennial visits of all childhood cancer survivors are expected to outgrow both current and future resources of our late effects outpatient clinic. In order to be able to develop risk-adapted follow-up programs it is deemed necessary to collect data on the medical and psychosocial problems encountered in the complete cohort of childhood cancer survivors previously treated at the EKZ/AMC. This study provides an overview of the problems in the first 976 survivors screened at our outpatient clinic. Since these 976 survivors do not represent the complete cohort of survivors no accurate relative risk estimations can be provided on the basis of these data. It is, however, possible to compare the observed problems in this cohort of childhood cancer survivors to some degree with those published previously. In this study cardiac dysfunction was registered in 4.5 % of the childhood cancer survivors. Approximately the same frequencies were observed by Stevens et al. [6] (3%) and by Sklar [8] (7.7% cardiopulmonary dysfunction). Oeffinger et al. [7], however, reported cardiovascular toxicity in 20 out of their cohort of 96 (20.8%). Tissue hypoplasia was observed in 20.7% of the survivors of our cohort. In contrast to the cardiac dysfunction this compares well with the group described by Oeffinger et al. [7] (24%) and not with the report by Stevens et al. [6] (1%). The frequency of mamma hypoplasia observed in the latter study [6] (1%) is in the same range as the 1.9% found here.

Several explanations can be given for the observed differences. Most importantly, not all studies use the same definitions and classification systems. This renders exact comparison between the different studies virtually impossible. This observation stresses the importance of the development of a uniform language in the field of late effects of childhood cancer. Another explanation for differences between late effects can be found in the different treatment schedules and philosophies used in the reporting countries, e.g., prophylactic cranial irradiation in leukaemia was abandoned in our country already early in the 1980's. Finally, an explanation can be found in different referral rates to the different reporting centres, also illustrated by the different distributions of primary diagnoses in the different studies [6-8]. There are several ways to overcome this problem. A multicenter survey study such as the Childhood Cancer Survivor Study (CCSS) [11,14,42] will equalise the influence of differential referral patterns. There are, however, some disadvantages to the questionnaire based studies. A major one is the response rate that was for the CCSS lower than thus far encountered at our outpatient clinic. As stated earlier, incompleteness of a cohort will influence the validity of relative risk estimates. In the last few years a group of paediatric oncologists representing all paediatric oncology centres in The Netherlands have been co-operating in the development of a uniform nation wide screening program. It is expected that with the establishment of the Dutch Childhood Oncology Group (DCOG) this initiative will lead to uniform screening for late effects after referral from active-treatment clinics to the late effects clinics. Uniform screening and central registration will allow for transfer from one late effects clinic to another when childhood cancer survivors are going to live independently. This possibility might be expected to lead to continuing high response rates.

Notwithstanding the differences between this study and several other reports, one thing is clearly evident. The vast majority of childhood cancer survivors suffer from at least one or some medical and /or psychosocial problems. It is our firm belief that it will also in the future remain necessary to provide adequate care and screening programs. The latter must, however, be developed on the basis of adequate relative risk estimates. As in the active treatment phase, where most childhood cancer patients are entered in clinical studies, it will for the coming decades remain necessary to perform additional, sometimes patient/survivor based research aimed at the improvement of the long-term care. Especially so since ever changing treatment strategies will lead to a changing pattern of observed late effects in the childhood cancer survivors. The organisation of our late effects clinic will continually allow for such additional studies. One major danger in performing these studies is the fact that childhood cancer survivors may get the feeling that visiting the outpatient clinic serves more the research purposes that the patient care. It is the responsibility of those organising the late effects programs to continually communicate to the survivors the results of these additional studies, and how these results will change the follow-up programs. Furthermore, it will remain very important to regularly ask the survivors about their own expectations [41].

From an economic point of view it might also be important to gain insight into costs of the screening procedures and referral patterns. In the near future a new version of PLEKsys will be implemented that also allows to register and analyse these type of data. It is the hope of the authors that both the completion of the epidemiological analyses that are currently in progress and some form of cost-benefit analysis will lead to an optimal largely evidence-based follow-up program with the lowest burden for not only the survivors' time, but also for the ever tight health-care budget in The Netherlands.

Appendix 1. Categories and subcategories of medical and psychosocial problems as registered in the PLEKsys database

Alopecia	**Fertility**	**Neurologic continued**	**Pulmonologic**
Cardiologic	low FSH/ LH	other motor disturbances	obstructive
cardiomyopathy	raised FSH/ LH	sensory disturbances	restrictive
arrythmias	small testes	nystagmus	diffusion abnormalities
valve abnormalities	oligo-/ amenorrhoea	other neurologic problems	**Dental problems**
wall movement abnormalities	oligo-/ azoospermia	**Ophthalmologic**	caries
Surgical	early menopause	abnormalities of vision	irregular teeth
splenectomy	**Gastro-enterologic**	cataract	xerostomia
nephrectomy left, right, bilateral	radiation enteritis	blindness left, right, bilateral	gingivitis
ovariectomy left, right, bilateral	hepatitis C	enucleation left, right, bilateral	**Subsequent neoplasm**
orchidectomy left, right, bilateral	bowel obstruction	**Orthopedic**	benign
Dermatological	malabsorption	scoliosis / kyfosis	malignant
scar due to extravasation of chemotherapy	**Hematologic**	amputaton / rotationplasty / prosthesis	**Urologic**
scar due to operation	**Ear, Nose and Throat**	anisomelia	hemorrhagic cystitis
multiple naevi	hearing disturbances	arthrosis	recurrent urinary tract infections
Endocrine	**Short stature**	lower back pain	urinary incontinence
panhypopituitarism	**Metabolic**	short vertebral column	anatomic abnormalities
growth hormone deficiency	dyslipidemia	short Achilles tendon	constitutional
secondary hypothyroidism	**Nephrologic**	other orthopedic problems	after therapy
primary hypothyroidism	tubulur dysfunction	**Pain**	**Fatigue**
precocious puberty	hypertension	headache	**Tissue hypoplasia**
diabetes mellitus	glomerular dysfunction	other pain	face
hypoparathyroidism	**Neurologic**	**Psychosocial / cognitive problems**	skull
hyperparathyroidism	epilepsy	learning difficulties	neck
adrenal insufficiency	neuropathy	memory disturbances	thorax
	hemiparesis / paralysis	concentration abnormalities	mamma
	dysarthria	social-emotional problems	abdomen
	ataxia		pelvis
	paresis of facial nerve		extremities
			skin
			Other

References

1. Coebergh JWW, Dijck JAAMv, Janssen-Heijnen MLG, Visser O. Childhood cancer in the Netherlands 1989-1997. 2000. Utrecht, Association of Comprehensive Cancer Centres.
2. Stiller CA, Draper GJ. The epidemiology of cancer in children. In: Voute PA, Kalifa C, Barrett A, editors. Cancer in children: clinical management. Oxford: Oxford University Press 1998, 1-20.
3. Cancer incidence and survival among children and adolescents: United States SEER program 1975-1995, National Cancer Institute, SEER Program. Ries LAG, Smith MA, Gurney JG et al., editors. [No. 99-4649] 1999, NIH Pub.
4. Heikens J. Childhood cancer and the price of cure; studies on late effects of childhood cancer treatment. Thesis, University of Amsterdam, 2000.
5. Meadows AT, Krejmas NL, Belasco JB. The medical cost of cure: sequelae in survivors of childhood cancer. In: Eys Jv, Sullivan MP, editors. Status of the curability of childhood cancer. New York: Raven Press 1980, 263-276.
6. Stevens MC, Mahler H, Parkes S. The health status of adult survivors of cancer in childhood. *Eur J Cancer* 1998, 34, 694-698.
7. Oeffinger KC, Eshelman DA, Tomlinson GE, Buchanan GR, Foster BM. Grading of late effects in young adult survivors of childhood cancer followed in an ambulatory adult setting. *Cancer* 2000, 88, 1687-1695.
8. Sklar CA. Overview of the effects of cancer therapies; the nature, scale and breath of the problem. *Acta Paediatr* 1999, Suppl, 433, 1-4.
9. Green DM, Hyland A, Chung CS, Zevon MA, Hall BC. Cancer and cardiac mortality among 15-year survivors of cancer diagnosed during childhood or adolescence. *J Clin Oncol* 1999, 17, 3207-3215.
10. Paulussen M, Ahrens S, Lehnert M et al. Second malignancies after ewing tumor treatment in 690 patients from a cooperative German/Austrian/Dutch study. *Ann Oncol* 2001, 12, 1619-1630.
11. Neglia JP, Friedman DL, Yasui Y et al. Second malignant neoplasms in five-year survivors of childhood cancer: childhood cancer survivor study. *J Natl Cancer Inst* 2001, 93, 618-629.
12. Garwicz S, Anderson H, Olsen JH et al. Second malignant neoplasms after cancer in childhood and adolescence: a population-based case-control study in the 5 Nordic countries. The Nordic Society for Pediatric Hematology and Oncology. The Association of the Nordic Cancer Registries. *Int J Cancer* 2000, 88, 672-678.
13. Moller TR, Garwicz S, Barlow L et al. Decreasing late mortality among five-year survivors of cancer in childhood and adolescence: a population-based study in the Nordic countries. *J Clin Oncol* 2001, 19, 3173-3181.
14. Mertens AC, Yasui Y, Neglia JP et al. Late mortality experience in five-year survivors of childhood and adolescent cancer: the Childhood Cancer Survivor Study. *J Clin Oncol* 2001, 19, 3163-3172.
15. Hawkins MM. Long term survival and cure after childhood cancer. *Arch Dis Child* 1989, 64, 798-807.
16. Robertson CM, Hawkins MM, Kingston JE. Late deaths and survival after childhood cancer: implications for cure. *Br Med J* 1994, 309, 162-166.
17. Hudson MM, Jones D, Boyett J, Sharp GB, Pui CH. Late mortality of long-term survivors of childhood cancer. *J Clin Oncol* 1997, 15, 2205-2213.
18. Relander T, Cavallin-Stahl E, Garwicz S, Olsson AM, Willen M. Gonadal and sexual function in men treated for childhood cancer. *Med Pediatr Oncol* 2000, 35, 52-63.
19. Kremer LC, van Dalen EC, Offringa M, Ottenkamp J, Voute PA. Anthracycline-induced clinical heart failure in a cohort of 607 children: long-term follow-up study. *J Clin Oncol* 2001, 19, 191-196.
20. Kremer LC, van Dalen EC, Offringa M, Voute PA. Frequency and risk factors of anthracycline-induced clinical heart failure in children: a systematic review. *Ann Oncol* 2002, 13, 503-512.
21. Lipshultz SE, Colan SD, Gelber RD, Perez-Atayde AR, Sallan SE, Sanders SP. Late cardiac effects of doxorubicin therapy for acute lymphoblastic leukemia in childhood. *N Engl J Med* 1991, 324, 808-815.
22. Lipshultz SE, Grenier MA, Colan SD. Doxorubicin-induced cardiomyopathy. *N Engl J Med* 1999, 340, 653-654.

23. Goorin AM, Chauvenet AR, Perez-Atayde AR, Cruz J, McKone R, Lipshultz SE. Initial conges-
tive heart failure, six to ten years after doxorubicin chemotherapy for childhood cancer. *J Pediatr* 1990, 116, 144-147.

24. Noorda EM, Somers R, van Leeuwen FE, Vulsma T, Behrendt H. Adult height and age at menar-
che in childhood cancer survivors. *Eur J Cancer* 2001, 37, 605-612.

25. Nysom K, Holm K, Michaelsen KF, Hertz H, Muller J, Molgaard C. Bone mass after treatment
for acute lymphoblastic leukemia in childhood. *J Clin Oncol* 1998, 16, 3752-3760.

26. Strauss AJ, Su JT, Dalton VM, Gelber RD, Sallan SE, Silverman LB. Bony morbidity in children
treated for acute lymphoblastic leukemia. *J Clin Oncol* 2001, 19, 3066-3072.

27. Sklar CA, Mertens AC, Walter A et al. Changes in body mass index and prevalence of overweight
in survivors of childhood acute lymphoblastic leukemia: role of cranial irradiation. *Med Pediatr
Oncol* 2000, 35, 91-95.

28. Lemerle J, Oberlin O, de Vathaire F, Pein F, Aubier F. Late and very late effects of therapy -
towards lifetime follow-up of cured patients. In: Voute PA, Kalifa C, Barrett A, editors. Cancer
in children - clinical management. Oxford: Oxford University Press 1998, 84-98.

29. Langeveld N, Ubbink M, Smets E. 'I don't have the energy': The experience of fatigue in young
adult survivors of childhood cancer. *EJON* 2002, 4, 20-28.

30. Kingma A, Rammeloo LA, Der Does-van den Berg A, Rekers-Mombarg L, Postma A. Academic career after treatment for acute lymphoblastic leukaemia. *Arch Dis Child* 2000, 82, 353-357.

31. Copeland DR, deMoor C, Moore BD, III, Ater JL. Neurocognitive development of children after
a cerebellar tumor in infancy: A longitudinal study. *J Clin Oncol* 1999, 17, 3476-3486.

32. Hobbie WL, Stuber M, Meeske K et al. Symptoms of posttraumatic stress in young adult survivors of childhood cancer. *J Clin Oncol* 2000, 18, 4060-4066.

33. Phipps S, Dunavant M, Srivastava DK, Bowman L, Mulhern RK. Cognitive and academic functioning in survivors of pediatric bone marrow transplantation. *J Clin Oncol* 2000, 18, 1004-1011.

34. Ris MD, Packer R, Goldwein J, Jones-Wallace D, Boyett JM. Intellectual outcome after reduced-
dose radiation therapy plus adjuvant chemotherapy for medulloblastoma: a Children's Cancer
Group study. *J Clin Oncol* 2001, 19, 3470-3476.

35. Palmer SL, Goloubeva O, Reddick WE et al. Patterns of intellectual development among survivors of pediatric medulloblastoma: a longitudinal analysis. *J Clin Oncol* 2001, 19, 2302-2308.

36. Langeveld NE, Grootenhuis MA, Voute PA, de Haan RJ, van den Bos C. Quality of life, self-
esteem and worries in young adult survivors of childhood cancer. *Psychooncology* 2002 (submit-
ted).

37. Langeveld NE, Ubbink MC, Last BF, Grootenhuis MA, Voute PA, de Haan RJ. Educational
achievement, employment and living situation in long-term young adult survivors of childhood
cancer in the Netherlands. *Psychooncology* 2002, 11, 1-13.

38. Neglia JP, Nesbit ME, Jr. Care and treatment of long-term survivors of childhood cancer. *Cancer*
1993, 71, 3386-3391.

39. Wallace WH, Blacklay A, Eiser C et al. Developing strategies for long term follow up of survivors of childhood cancer. *BMJ* 2001, 323, 271-274.

40. Oeffinger KC, Eshelman DA, Tomlinson GE, Buchanan GR. Programs for adult survivors of
childhood cancer. *J Clin Oncol* 1998, 16, 2864-2867.

41. Eiser C, Levitt G, Leiper A, Havermans T , Donovan C. Clinic audit for long-term survivors of
childhood cancer. *Arch Dis Child* 1996, 75, 405-409.

42. Kadan-Lottick NS, Robison LL, Gurney JG et al. Childhood cancer survivors' knowledge about
their past diagnosis and treatment: childhood cancer survivor study. *JAMA* 2002, 287, 1832-
1839.

Chapter 3

Quality of life in young adult survivors of childhood cancer: a literature review

Abstract

Purpose: In recent years the necessity of measuring quality of life in childhood cancer survivors has been stressed. This paper gives an overview of the results of studies into the quality of life (QL) of young adult survivors of childhood cancer and suggest areas for future research.

Methods: A literature search of studies published up to 2001 was conducted using the data bases of MEDLINE, CINAHL , EMBASE and PsychINFO.

Results: The review located 30 empirical studies published up to 2001. The results are described in terms of the following quality-of-life dimensions: physical functioning (general health), psychological functioning (overall emotional functioning, depression and anxiety, self-esteem), social functioning (education, employment, insurance, living situation, marital status and family), and sexual functioning. Factors related to survivors' QL are reported: demographics and illness- and treatment related variables. Although the literature yields some inconsistent findings, a number of clear trends can be identified: a) most survivors reported to be in good health, with the exception of some bone tumour survivors; b) most survivors function well psychologically; c) survivors of CNS tumours and survivors of acute lymphoblastic leukaemia (ALL) are at risk for educational deficits; d) job discrimination, difficulties in obtaining work and problems in obtaining health and life insurance's were reported; e) survivors have lower rates of marriage and parenthood; f) survivors worry about their reproductive capacity and/or about future health problems their children might experience as a result of their cancer history.

Conclusion: There is a need for methodological studies that measure QL among survivors of childhood cancer more precisely by taking into account the effects of the severity of the cancer and the long-term impact of different treatments. Additional data are needed to help us understand the needs of survivors and to identify those subgroups of survivors who are at greatest risk for the adverse sequelae of the disease and its treatment.

Introduction

As a result of more effective treatment, improved supportive care and centralisation of care, the long-term survival rate of childhood cancer patients has risen dramatically during the past few decades. More than 70% of children newly diagnosed with acute lymphocytic leukaemia will be in continuous remissions 5 years following their initial diagnosis, and the majority of these patients are probably cured of their disease. Survival has also increased for children with solid tumours: 93% of children with Hodgkin's disease, 84% of children with Wilms' tumour and 73% of children with non-Hodgkin's lymphoma will be alive 5 years after diagnosis [1].

The same treatments as have enabled long-term survival, however, can also cause potentially debilitating deficits, ranging from disruptions in day-to-day activities to such late effects as second primary cancers [2-4]. While numerous long-term physical effects of childhood cancer have been documented, the impact of such sequelae on patients' quality of life (QL) is much less understood. Although a growing number of studies have documented the considerable impact of cancer diagnosis and treatment on QL in short-term survivors, less attention has focused on QL in long-term young adult survivors, partly because the rise in survival rates is relatively recent.

It is evident that long-term effects in young adults may differ from those experienced in childhood or adolescence. New issues may come up that were not of concern earlier on. For example, worries about fertility and health of offspring may not emerge until the survivor has reached a certain age and is in a stable relationship and both partners would really want to have children. Some of the late physical effects of childhood cancer treatment, such as those resulting from the cardiotoxic effects of some chemotherapeutic agents, are only just being identified, and how these sequelae may affect the survivors' QL is not known.

The research on QL in young adult survivors of childhood cancer is reviewed in this paper. The purpose of this article is to give investigators and other persons involved in childhood cancer care an overview of the research that has been conducted in this field. On the basis of the literature, limitations of the studies and methodological difficulties are described. Finally, suggestions for future research are given.

The concept of quality of life

Assessment of QL is complicated by the fact that there is no universally accepted definition for it. In the past, most researchers measured only one dimension, such as physical function, economic concern, or sexual function. More recently, researchers have attempted further definition of QL. The World Health Organization defines QL as "individuals' perceptions of their position in life in the context of the culture and value system in which they live and in relation to their goals, standards, and concerns" [5]. The definition includes six broad domains: physical health, psychological state, levels of independence, social relationships, environmental features, and spiritual concerns. The importance of this definition to childhood cancer survivors lies in the inclusion of both emotional and social dimensions of health in addition to physical aspects. While many survivors have no physical evidence of disease and appear to have made full recoveries, others have to come to terms with the chronic, debilitating, or delayed effects of therapy. All remain at risk of the development of late sequelae of the former disease and/or treatment and of second malignancies. Furthermore, in most cases the life-threatening experience of cancer is never forgotten. In many ways, survival enhances appreciation for life, while at the same time reminding survivors of their vulnerability. The metaphor of the Damocles syndrome illustrates this dichotomy and the way individual survivors interpret this metaphor for life will influence the quality of their survival [6].

Methods

A literature search of studies published up to 2001 was conducted using the data bases of MEDLINE (National Library of Medicine), CINAHL (Cumulative Index to Nursing and Allied Health Literature), EMBASE and PsychINFO. The keywords 'childhood cancer', '(long-term) survivors', and 'late effects' were combined with dimensions that are often included as components of QL, including psychological/social adjustment, employment/health insurance, schools/ learning, and quality of life/health status. Relevant articles were then hand-searched for further pertinent references. Studies published in English were included in the review. This review has been performed according to the methodological criteria defined by Eiser and colleagues [7] for the inclusion of studies in the field of psychosocial paediatric oncology. These standards are: 1) well-validated and reliable measures, 2) well-matched control group, or comparison with culturally appropriate measurement norms, 3) information about demographics and about illness and treatment factors (at least cancer diagnosis and time since diagnosis), 4) respondent rate, and 5) use of appropriate rigorous statistical tests. Additional selection criteria applied included: 6) survivors as the primary source of QL information, either by means of interviews or by completion of self-report questionnaires (studies with no more than 20% proxies as primary source of information were also included in this review), 7) original diagnoses of survivors made before they were 20 years of age, and 8) at least 5 year's survival after completion of therapy. Some studies, however, included survivors of 5 or more years after completion of therapy along with respondents who were closer to completion of therapy. We decided to include these studies as well because in most cases the mean or median time since completion of therapy ranged from 6 to 15 years. In addition, there is no consensus in the paediatric oncology literature about the definition of a survivor. Some authors define a survivor as a child or adolescent who has been disease-free for at least 5 years, while others use disease-free survival for 2 years or more as their criterion. This may be partly due to the different survival perspectives for the different diagnoses in childhood cancer. We intended to limit this review to studies of survivors who were at least 18 years old at the time of investigation. However, a number of studies included survivors both under and over the age of 18 years. Studies of QL that included patients with very wide age ranges were excluded on the grounds that in these studies it is not possible to distinguish the impact of cancer on children from that on older adults. Studies in which results for the long-term survivors, or time since completion of therapy were reported separately are included. In most study reports, however, this was not the case.

Initially, the studies featured in this review were selected by two reviewers on the basis of the above methodological criteria. However, we found that in most studies survivors' social functioning (e.g. education, employment) was not measured with the aid of standardised, well-validated instruments. Because we did not want to exclude the social aspects of survivors' QL we decided to include these studies in the review as well, while being aware of the methodological limitations. At the same time, these limitations meant that there was no possibility of doing a systematic review.

A total of 30 empirical studies that met the inclusion criteria were found. We found the results of one study in two different journals, and we have combined these findings in our review [8,9]. The studies are summarised in Table 1. In this table, the following information is provided for each study: a) the first author and year of publication; b) type of cancer; c) number and sex of survivors; d) age at evaluation; e) age at diagnosis and time since completion of therapy; f) number, sex and age of subjects in control group; g) instruments/measures used; and h) results/outcome. These parameters are reviewed and summarised in the following five sections. The first section describes physical functioning, including QL and general health. The second section summarises the results relating to psychological functioning: overall emotional functioning, depression and anxiety, and self-esteem. No studies were found about the cognitive or neuropsychological aspects of psychological functioning in childhood cancer survivors. The third section describes social functioning, including education, employment, insurance, living situation, marital status and family. Sexual functioning is the topic of the fourth section, and in the fifth section factors related to survivors' functioning are summarised: demographics, and illness- and treatment related factors.

Table 1. Summary of selected studies in young adult survivors of childhood cancer

Study (yr)	Type of cancer	Number and sex of survivors	Age at evaluation (yr)	Age at diagnosis and time since completion of therapy (yr)	Number, sex and age of control group	Instruments/measures	Results/outcome
Holmes et al. [17] (1986)	Mixed	100 58% men 42% women	21 + · mean age 33 median age 32	range 0 - 19 5+	100 sex-matched siblings mean age 33 median age 32	structured interview in person or by telephone about life and health insurance	Survivors had significantly more difficulty in obtaining life insurance and in obtaining health insurance because of health reasons. Survivors were significantly less likely than siblings to be covered by health insurance
Lansky et al. [20] (1986)	mixed (excluding CNS tumors)	39 49% men 51% women	mean age 23 age range 16 - 33	mean 13 range 10 - 18 5+ since diagnosis mean 7 range 0 - 18	siblings	structured interview in person or by telephone about family demographics, past and present medical problems, academic and career achievements, disease impact, issues related to separation, overall psychological adjustment	No significant difference in marital status, living arrangements, academic and career attainment. A higher prevalence rate in episodes of treated depression, alcoholism, and/or suicide attempts in survivors compared with that in the general population. Almost half reported that academic plans were altered, and 38% had made changes in their career goals because of the illness
Teta et al. [23] (1986)	mixed	450 48% men 52% women	21 +	< 19 5+ since diagnosis	587 sex-matched siblings 47% men 53% women age >19	structured interview in person or by telephone: Addendum at NCI questionnaire about depression, difficulties in obtaining entrance to armed forces or access to educational and employment opportunities, health and life insurance	No significant difference in depression between survivors and siblings, frequency was similar to the general population. No differences in reported frequencies of suicide attempts, running away or psychiatric hospitalisations. Survivors experienced significantly more jobdiscrimination, more rejection from the armed forces and had more difficulty obtaining health and life insurance than their siblings. Depression was not associated with: age at diagnosis, year of diagnosis and type of treatment
Teeter et al. [24] (1987)	mixed	263 52% men 48% women	21 + mean age 33 age range 23 - 54	< 20 5+ since diagnosis	369 sex-matched siblings 47% men 53% women mean age 33	structured interview in person or by telephone: Part of NCI questionnaire including decisions about marriage and family	Survivors were less likely to marry. More survivors than controls were not married for health reasons. More survivors than controls reported that they never have been pregnant and having no

Study	Tumor type	N / sex	Age at follow-up	Age / time since diagnosis	Controls	Assessment	Results
					age range 20 - 59		offspring. No significant difference was found between frequency of birth defects among the offspring of both survivors and controls
Wasserman et al. [38] (1987)	Hodgkin's Disease	40 55% men 45% women	mean age 25 age range 10 -38	< 20 mean 13 range 12 - 19 / 5+ since diagnosis mean 12 range 7 - 19	no control group, population norms available	structured interview in person about perceptions about their cancer, reactions of family and friends, risk-taking behavior, perceived benefits, education, employment, current medical problems; standardised questionnaire: Diagnostic and Statistical Manual of Mental Disorders	Survivors educational levels exceeded those expected in sex-age- and state-matched populations. Overall proportions of marriage and divorce did not differ from general population statistics. Male survivors had a higher rate of divorce compared with age- and race-specific statistics. Frequency of psychiatric diagnoses were not different from that found in the community
Kelaghan et al. [19] (1988)	mixed	2283 50% men 50% women	> 21 mean age 31 age range 21 - 55	< 20 range <5 - 19 / 5+ since diagnosis	3261 sex-matched siblings 49% men 51% women mean age 33 age range 19 - 70	Structured interview in person or by telephone about demographic characteristics, medical and reproductive history, social characteristics, including highest educational level achieved	No significant differences in educational achievement were found for survivors of non-CNS cancers. Survivors of CNS tumors were significantly less likely than controls to complete eight grades of school or, if they completed high school, to enter college. Educational deficit of brain tumour survivors was especially striking for tumors of the ventricles or cerebral hemispheres and the deficit was more severe for those who were younger at diagnosis and those treated with radiation therapy than by surgery alone
Tebbi et al. [22] (1989)	mixed (excluding CNS tumors)	40 40% men 60% women	mean age 26 age range 18 - 35	mean 16 range 13 - 19 / 5+ mean 10 since diagnosis	40 healthy sex- and age-matched controls 38% men 63% women mean age 26 age range 18 - 35	semi-structured interview by telephone with standardised questionnaires: The Rand Health Insurance Study Functional Limitations Battery (FLB), Physical Abilities Battery (PAB), vocational, insurance, social status, The	No differences with regard to overall general well-being, although survivors were more concerned about their health and reported lower general spirits. Survivor's health limited their ability to engage in vigorous activities. Survivors reported disease-related discrimination in hiring, induction into military service, and obtaining health, life, and disability insurance. Survivors did not differ with

36

Study	Tumor type	Sample size / sex	Age	Age at diagnosis / follow-up	Control group	Measure	Results
						Rand Health Insurance study General Well-Being measure	respect to employment status but they reported a higher average income than controls
Byrne et al. [11] (1989)	mixed	2170 sex not given	> 21 mean age 31	< 20 5+ since diagnosis	3138 sex-matched siblings sex not given mean age 33	structured interview in person or by telephone about demographic characteristics, personal medical history, marriage, divorce, pregnancies, offspring, fertility	Survivors were less likely to be married. Men treated for CNS tumors were the most serious affected. Not only were they less likely to be married, their first marriages were shorter and they were older at first marriage. Increased divorce rate in male survivors of retinoblastoma
Meadows et al. [21] (1989)	mixed	95 53% men 47% women	> 18 mean age 24 median age 22 age range 18 - 35	< 16 mean 6 5+	number of siblings not given median age 25	structured interview by telephone about educational achievement, occupational status, interpersonal relation-ships, including marital status and progeny, benefits and insurance concerns, medical and health behaviours	Good overall functioning. No difference in the amount of education between survivors and their siblings. Siblings were significantly more likely to be married. Many survivors worried about recurrence of cancer. History of cancer sometimes affected their relationships. Age at diagnosis and the type of treatment was not related to education level
Mäkipernaa [39] (1989)	solid tumors (excluding CNS tumors)	94 51% men 49% women	median age 23 age range 11 - 15	median 3 range 0 - 18	no control group, population norms available	semi-structured interview in person about education, occupation, social security, interests, marital status, disease-related opinions	Most survivors had good adjustment, some are at risk of developing emotional and social problems. Education level was similar, or slightly above population level. Some males were rejected for military service because of the history of cancer. Fewer of the females and as many of the males were married as in the general population
Zevon et al. [31] (1990)	ALL	46 52% men 48% women	mean age 23 age range 18 - 34	< 20 mean 7 range 2 - 18 5+ since diagnosis mean 15	population norms available control group with HD + NHL	standardised question-naires in person or by mail: Multi-dimensional Personality Questionnaire (MPQ) Well-Being and Stress Reaction Scales, Minnesota Satisfaction Questionnaire-Short Form	ALL survivors appeared to be well-adjusted. Female survivors, however, had an increased tendency to experience anxiety in stressful situations. ALL survivors were marrying at a somewhat lower rate than the overall population. Vocational satisfaction did not differ from population norms. Vocational discrimination did not appear to be a

Study		N / Sex	Age	Age at diagnosis / time since	Control group	Measures	Results
Green et al. [15] (1991)	mixed	227 54% men 46% women	median age 27 age range 18 - 44	< 20 median 11 range 1 - 19 5+ since diagnosis	no control group, population norms available	(MSQ), Long-Term Follow-Up Questionnaire (LFQ) (medical, employment, marital, and family history), occupational status self-report questionnaire in person or by mail about marital status, employment history, current occupation and job duties, health and life insurance status, reproductive history, family history	problem. Cranial irradiation was negatively associated with well-being. The percentage of employed survivors was not different from US norms. Percentages of life and health insurances were lower than US population. The percentages of married men and women were significantly lower than US norms, especially women aged 20-24. Women aged 35-44 had a significantly higher frequency of divorce compared with age-specific group norms. Male gender and age at study was positively associated with employment. Diagnosis, age at diagnosis and disease recurrence were not related to employment, marriage, divorce and insurance
Hays et al. [16] (1992)	mixed	219 sex not given	30+	< 19 2+	190 sex-matched siblings or friends	structured interview by telephone or mail about insurance coverage, demographic questions on race and ethnicity, occupation, education, employment, income	No differences in education, employment and insurance between non-CNS survivors and controls. Survivors of CNS tumors had limited educational achievements and lower rates of marriage and parenthood
Gray et al. [8] (1992) Gray et al. [9] (1992)	mixed	62 65% men 35% women	> 18 mean age 26 age range 18 - 37	< 18 mean 11 range 1 - 18 2+ mean 15 range 2 - 33	51 healthy age-matched peers 45% men 55% women mean age 26	standardised question-naires in person: Profile of Mood States, Desirability of Control Scale, Control Belief Scale, Locus of Control Scale, Rosenberg Self-Esteem Scale, Impact of Event Scale, projective story-telling technique, screening questionnaire (demographic factors,	Survivors were similar to their peers in overall psychological adaptation, both within normal ranges. Survivors reported more positive affect, less negative affect, higher intimacy motivation, more perceived personal control and greater satisfaction with control in life situations. Survivors, especially CNS survivors, were more likely to have repeated school grades. Further, survivors worried more about issues of fertility and expressed more dissatis-

Study	Diagnosis	Survivors	Age	Time since diagnosis / age at diagnosis	Controls	Method	Results
Nicholson et al. [35] (1992)	osteosarcoma + Ewing's sarcoma	111 50% men 50% women	> 21 mean age 33 age range 21 - 51	< 20 mean 15 range 3 - 19 5+ since diagnosis mean 18	151 sex- and age-matched siblings 44% men 56% women mean age 33 age range 21 - 66	structured interview in person or by telephone about health status, activities of daily living, education, employment and disability, marriage, fertility, pregnancy, health of their offspring	faction with important relationships. No effects of time since illness, age at diagnosis, presence of recurrence and report of disability. Osteosarcoma survivors were more likely than their siblings to perceive their health as poor. Survivors were more likely than controls to have some difficulty climbing stairs and to have had employment disability. Marriage rate, fertility, employment status and annual income were similar. Amputation status was not associated with health perception
Haupt et al. [26] (1994)	ALL	593 51% men 49% women	> 18 mean age 23 age range 18 - 33	< 20 median 10 range 0 - 20 2+ since diagnosis	409 sex-matched siblings 46% men 54% women mean age 25 age range 18 - 42	structured interview by telephone about education (highest level of schooling, average grades during high school, enrollment into special programs	On average, survivors had lower grades, were more likely to enter a special education or a learning disabled program and spent longer time in these programs. Survivors were at higher risk of missing school for long periods and/or repeating 1 year of school. Most survivors have rates of high school graduation, college entry, and college graduation that are similar to their siblings. Survivors treated with 24 Gy of CRT and those diagnosed at a preschool age were at higher risk for poor educational performance
Evans & Radford [14] (1995)	mixed	48 54% men 46% women	mean age 20 age range 16 - 30	information about age at diagnosis not given < 5 (16 survivors) 5 + (32 survivors)	38 siblings mean age 21 age range 16 - 30	unstructured interview in person about their experiences of cancer; structured interview: questions about their illness and current lifestyle; standardised questionnaire: Oxford Psychologists Press adult self-esteem questionnaire	No significant difference in their educational achievements, employment status and salary earned, driving test achievements, establishing relationships, partaking in societies and competitive sports. Survivors were less likely to go on to higher education. Survivors overall self-esteem was as high as their siblings
Jacobson Vann et al. [18] (1995)	mixed	187 47% men 53% women	age range 19 - 39	< 19 5+ since diagnosis	108 siblings 43% men 57% women	questionnaire by mail about health insurance	Survivors were found to be more likely to be denied health insurance because of their cancer history and related medical

Study	Diagnosis	N / Sex	Age	Age at diagnosis / Follow-up (years)	Controls	Assessment	Results
					age range 19 - 39		history than their siblings. Survivors also had health insurance policies that excluded care for pre-existing medical conditions more often. Survivors reported more problems obtaining health insurance coverage, were more likely to be covered by their parents' health insurance policies and had been turned down for a job more often because of their cancer history
Apajasalo et al. [10] (1996)	mixed (excluding CNS tumors)	168 63% men 37% women	mean age 23 age range 16 - 35	information about age at diagnosis not given 1+ median 12	129 persons general population 47% men 53% women age range 17 - 35	standardised questionnaire by mail: 15-dimensional questionnaire (15D) (mobility, vision, hearing, breathing, sleeping, eating, speech, elimination, usual activities, mental function, discomfort and symptoms, depression, distress, vitality and sexual activity)	Survivors QL was significantly better than that of controls. Both the survivors and the controls reported good levels of physical dimensions, sensory dimensions, usual activities and mental function. Although emotional dimensions (depression, distress, vitality, sleeping and discomfort) were less satisfactory in both groups, survivors reported less problems than the controls. Younger survivors reported a better QL. BMT survivors reported a slightly lower QL. Type of cancer, follow-up time and gender were not associated with QL
Puukko et al. [29] (1997)	Acute Leukaemia	30 100% women	> 16 mean age 20	mean 9 1+ mean 8	50 healthy age-matched controls 100% women mean age 20	self-report questionnaire, semi-structured interview, psychiatric evaluation, psychological tests in person about sexual attitudes, fears and behaviours, family and peer relationships, sexual experiences, health and illness concerns, ideals and expectations from life	The age at initiation of dating and sexual activity, the frequency of sexual intercourse, and opinions on sexual behavior were similar. Survivors differed significantly from controls with regard to inner sexuality: images of sexuality more restrictive, attitudes (concerning sexual pleasure) more negative. Sexual identity less often feminine and more often infantile in survivors. Sexual identity was not associated with age at study, age at diagnosis, type of treatment and follow-up time
Moe et al. [28] (1997)	ALL (without CRT)	94 55% men 45% women	not given	mean 5 15+	90 sex-matched siblings/cousins 59% men 41% women	standardised questionnaires by mail: SF-12 (physical and mental health), General	No statistical difference with respect to physical and mental health and QL. The somatization score on the GHQ involving items closely related to fatigue

Reference	Diagnosis	N / sex	Age	Age at / time since diagnosis	Control group	Instruments	Results
						Health Questionnaire (depression, anxiety, fatigue, social dysfunction), Eysenck's short scale of the EPQ-R (possible late effects on personality); author-developed questionnaire: demographic data, number of offspring, learning problems, level of athletic performance, education and work status	demonstrated a significantly higher score for the ALL survivors. No significant differences with regard to performances issues, such as academic skills, level of education, work status and level of physical exercise. Male survivors had significant fewer offspring than their male controls
Zeltzer et al. [30] (1997)	ALL	580 51% men 49% women	mean age 23 median age 22 age range 18 - 33	< 20 2+ since diagnosis	396 sex-matched siblings 46% men 54% women mean age 25 median age 25 age range 18 - 41	structured interview by telephone about education, marital status, employment status, health, fertility, offspring, risk behaviours; standardised questionnaire: Profile of Mood States (POMS) (tension, anxiety, depression, anger, confusion, vigor, fatigue)	Marital differences between survivors and controls were not significant, however, older survivors were more likely to never have married. Survivors were more likely to be unemployed or working less than half-time. Survivors had a greater negative mood, reported more tension, depression, anger and confusion than controls, however, scores were not as high as those found in a psychiatric sample. No differences on the vigor and fatigue subscale scores. Female, minority, and unemployed survivors reported highest total mood disturbance
Elkin et al. [13] (1997)	mixed	161 53% men 47% women	> 15 median age 19 range 15 - 31	median 10 range 0 - 21 2+ median 7 range 2 - 15	no control group, population norms available	standardised questionnaire in person: Symptom Checklist-90 Revised (SCL-90-R) (somatization, obsessive-compulsive, interpersonal sensitivity, depression, anxiety, hostility, phobic anxiety, paranoid ideation, psychoticism, distress, cosmetic and functional impairments)	Mean scores on all subscales of SCL-90-R were lower than those of the standardisation sample, distributions of scores on the anxiety, psychotism, Global severity index, and positive symptom total scores were significantly below normative values. Survivors appear significantly healthier than age- and gender matched norms for the general population. Older age, disease relapse and functional impairment were risk factors for maladjustment. Not associated were diagnosis, age at

Study	Diagnosis	Number	Age at study	Time since diagnosis	Controls	Measures	Results
Novakovic et al. [36] (1997)	Ewing's sarcoma family tumors	89 54% men 46% women	mean age 29 age range 10 - 48	mean 15 range 4 - 34 mean 13 range 2 - 29	97 sex- and age-matched siblings 47% men 53% women mean age 31 age range 10 - 57	questionnaire by mail about education, job history, marital status, fertility, health status, diet, exercise habits, health insurance, health care needs; standardised instrument: Karnofsky performance scale (current functional status)	associated were diagnosis, age at diagnosis, type of treatment, cosmetic status, gender and socioeconomic status. No difference in educational achievement. Survivors were less likely to be employed full-time, to be married and to have children. Survivors were more likely to have divorced than their siblings. No difference in self-rating of health status or in health care insurance status, but more problems in getting job-related health insurance. Functional status was adversely affected in survivors, they scored significantly worse than sibling controls. Having children was not related to treatment-related factors. Marriage was not associated with the treatment protocol and body irradiation
Felder-Puig et al. [34] (1998)	osteosarcoma + Ewing's sarcoma	60 43% men 57% women	mean age 24 age range 15 - 30	mean 15 range 0 – 25 1+ since diagnosis mean 8 range 1 - 21	no control group, population norms available	standardised questionnaires in person: Questionnaire on Subjective Well-Being (positive attitudes towards life, depressive mood, joy of life), State-Trait-Anxiety Inventory (trait-anxiety), Frankfurt Self-Concept Scales, Questionnaire on Life Goals and Satisfaction with Life; semi-structured interview about socioeconomic issues, life-style, problems and limitations due to disease and its consequences, overall quality of life	Survivors did not show a higher rate of serious personality disturbances or psychosocial problems then controls. Many, however, dealt with problems such as restricted mobility, catching up at school, changing jobs or job orientation. Survivors appeared to be married at a lower rate and seemed to live at home longer. Levels of education and income were similar. Survivors diagnosed in adolescence had more problems (especially social well-being) than those diagnosed in childhood or early adulthood. Physical or functional sequelae and disease-related variables were not associated with psychosocial adjustment
Dolgin et al. [12] (1999)	mixed	64 47% men 53% women	> 18 mean age 24 age range 18 - 35	< 18 mean 12 range 1 - 17	51 age- and sex-matched controls 53% men 47% women	structured interview in person about level of functioning and achievement in the domains of	No differences in education, employment, marital status and relationships with significant others. No evidence of increased psychological

Study	Diagnosis	N / sex	Age	Time since diagnosis	Control	Measures	Results
				3+ mean 10 range 3 - 21	mean age 23 age range 18 - 32	education, employment, military service, social/family status, health status; standardised questionnaire: Mental Health Inventory (anxiety, depression, loss of control, general positive feeling, positive relationships)	impairment or pathology. Survivors experienced military recruitment difficulties, lower income levels and higher rates of workplace rejection. Almost half reported feelings that their illness experience had impaired their achievement in education, social and family goals
Rauck et al. [25] (1999)	mixed	10425 54% men 46% women	median age 26 range 15 - 48	< 21 median 7 range 0 - 21 5+ since diagnosis	no control group, population norms available	questionnaire by mail about marital status	Survivors, especially females and whites, were less likely to have ever married, but, once married, were less likely to divorce/separate. Black survivors were generally found to be more likely to have married, with males and blacks more likely to divorce/separate once married. Survivors of CNS tumors, particularly males, were less likely to have ever married and more likely to divorce/separate compared to those with other diagnoses and the general US population
Veenstra et al. [37] (2000)	bone tumor	33 55% men 45% women	> 16 mean age 25 range 16 - 50	Information about age at diagnosis not given 1+ mean 6 range 1 - 11	no control group, population norms available	standardised questionnaires by mail: 2 items EORTC-QLQ-C30, SF-36, shortened version Social Support List-Interactions and Social Support List-Discrepancies, items from EORTC QLQ-BR23, physical functioning and prosthesis, negative and positive effects of surgery	Survivors physical functioning was poorer than that of healthy peers but better than chronically ill patients. Levels of psychosocial functioning, general QL and social support were highly comparable with healthy peers
Kingma et al. [32] (2000)	ALL (with CRT)	94 43% men 57% women	median age 20 range 15 - 32	median 5 range 1 - 15 mean 15	134 siblings 49% men 51% women median age 19 age range 14 - 32	questionnaire by mail about school career	Significantly more survivors than siblings were placed in special educational programmes. Survivors had also a lower level of secondary education. Younger age at diagnosis was negatively related to educational level.

| Mackie et al. [33] (2000) | ALL + Wilms tumour | 102 56% men 44% women | mean age 26 range 20 - 31 | < 16 mean 5 range 0 - 15

5+ mean 16 range 5 - 29 | 102 healthy age- and sex-matched controls 56% men 45% women mean age 26 age range 20 - 31 | standardised questionnaires in person; Schedule for Affective Disorder and Schizophrenia lifetime (SADS-L) with DSM-III-R, Adult Personality Functioning Assessment (APFA), Raven's standard progressive matrices | negatively related to educational level. Gender and cranial irradiation dose were not associated with educational level

No increased rates of psychiatric disorder for survivors. Mean scores of survivors were significantly higher (indicating poorer functioning) than controls for love/sex relationships, friendships, non-specific social contacts and day-to-day coping. Mean overall work and educational performance scores did not differ between groups. Poor close relationship was related to longer duration of treatment and more recent illness, while age at diagnosis was not associated |

Abbreviations: yr: years; CNS: central nervous system; ALL: Acute Lymphoblastic Leukaemia; CRT: Cranial Irradiation; HD: Hodgkin's Disease; NHL: non-Hodgkin's lymphoma

Results

Of the 30 studies, 17 involved survivors of different cancers and did not attempt to distinguish between diagnostic groups in terms of outcome [8,10-25]. Three of these studies excluded survivors of a CNS tumour [10,20,22]. Six studies focused specifically on leukaemia survivors [26,28-32], and Mackie and colleagues [33] included survivors who had been treated for acute lymphoblastic leukaemia (ALL) and Wilms' tumour. Four studies examined survivors treated for a bone tumour [34-37]; one study investigated Hodgkin's disease survivors [38]; and one study was found in which survivors treated for solid tumours, except for CNS tumors, were investigated [39].

The majority of the studies in this review were conducted in the United States [8,11,13,15-26,30,31,35,36,38,40]. Three studies were conducted in Finland [10,29,39], 2 in the United Kingdom [14,33], 2 in the Netherlands [32,37], 1 in Norway [28], 1 in Austria [34], and 1 in Israel [12]. Sample sizes varied from 30 [29] to 10425 [25]. Survivors differed in age at the time of evaluation (range from 10^1 [36,38] to 55 years [19], age at diagnosis, and time since completion of therapy. Twelve investigators used time since diagnosis as a criterion. Most investigators (n = 22) compared the results in survivors with those in sex- and age-matched siblings, peers or healthy controls. Seven studies included comparison with population norms [13,15,25,34,37-39], and 1 study included both population norms and a control group of survivors with a different cancer diagnosis [31]. The instruments used in most of the studies were a mixture of standardised questionnaires and tests (see Table 2). In the remainder the instruments were mostly newly developed questionnaires with no information given on reliability and validity, or authors used less highly structured interviews.

Physical functioning

Four investigations asked survivors for a general evaluation of their health. The majority of the survivors (89%) in the study by Meadows and colleagues [21] reported being in good to excellent health. Nicholson and colleagues [35] investigated 111 bone tumour survivors, and 80% of osteosarcoma and 100% of Ewing's sarcoma survivors classified their health as good or excellent. Similar findings of apparently good health when compared to siblings were reported by Novakovic and colleagues [36], who studied 89 survivors of Ewing's sarcoma family tumors. However, the osteosarcoma suvivors in the study by Nicholson and colleagues [35] were more likely than their siblings to perceive their health as fair or poor; this was neither explained by an excess of chronic health condition nor related to amputation status. When Dolgin and colleagues [12] asked participants to rate their current health status on a five point scale, survivors rated their health as poorer than controls.

---

1 As mentioned in the methods section, a number of studies included both survivors under the age of 18 years as survivors above the age of 18 years

Table 2. Standardised instruments used in quality of life studies of survivors of childhood cancer

Instrument	Studies using instrument
Physical functioning	
15 D (15-dimensional questionnaire)*	Apajasalo et al., 1996
Items EORTC QLQ-BR23	Veenstra et al., 2000
Items EORTC QLQ-C30	Veenstra et al., 2000
General Health Questionnaire*	Moe et al., 1997
Karnofsky performance scale	Novakovic et al., 1997
Physical Abilities Battery (PAB)	Tebbi et al., 1989
Profile of Mood States (POMS) fatigue subscale*	Zeltzer et al., 1997
SF-12*	Moe et al., 1997
SF-36	Veenstra et al., 2000
The Rand Health Insurance Study Functional Limitations Battery (FLB)	Tebbi et al., 1989
The Rand Health Insurance Study General Well-Being measure*	Tebbi et al., 1989
Psychological functioning	
Overall emotional functioning	
15 D (15-dimensional questionnaire)*	Apajasalo et al., 1996
General Health Questionnaire*	Moe et al., 1997
Mental Health Inventory	Dolgin et al., 1999
Multi-dimensional Personality Questionnaire (MPQ) Well-Being and Stress Reaction Scales	Zevon et al., 1990
Profile of Mood States (POMS)*	Gray et al., 1992; Zeltzer et al., 1997
Questionnaire on Subjective Well-Being	Felder-Puig et al., 1998
SF-12*	Moe et al., 1997
Symptom Checklist-90 Revised	Elkin et al., 1997
The Rand Health Insurance Study General Well-Being measure*	Tebbi et al., 1989
Depression and anxiety	
Diagnostic and Statistical Manual of Mental Disorders (DSM)	Wasserman et al., 1987
General Health Questionnaire*	Moe et al., 1997
Mental Health Inventory*	Dolgin et al., 1999
Profile of Mood States (POMS)*	Zeltzer et al., 1997
Schedule for Affective Disorder and Schizophrenia lifetime (SADS-L)	Teta et al., 1986; Mackie et al. 2000
State-Trait-Anxiety Inventory	Felder-Puig et al., 1998
Symptom Checklist-90 Revised	Elkin et al., 1997
Self-esteem	
Frankfurt Self-Concept Scales	Felder-Puig et al., 1998
Rosenberg Self-Esteem Scale	Gray et al., 1992
Oxford Psychologists Press adult self-esteem Questionnaire	Evans & Radford, 1995
Other	
Control Belief Scale	Gray et al., 1992
Desirability of Control Scale	Gray et al., 1992
Eyseneck's short scale of the EPQ-R	Moe et al., 1997
Impact of Event Scale	Gray et al., 1992
Locus of Control Scale	Gray et al., 1992
Minnesota Satisfaction Questionnaire-Short Form (MSQ)	Zevon et al., 1990
Questionnaire on Life Goals and Satisfaction with Life	Felder-Puig et al. 1998
Raven's standards progressive matrices	Mackie et al., 2000
Social functioning	
Adult Personality Functioning	Mackie et al., 2000
Long-term Follow-up Questionnaire (LFQ)	Zevon et al., 1990
Social Support List-Interactions and Social Support List-Discrepancies	Veenstra et al., 2000
Sexual functioning	
No instruments	

* Questionnaire consisting of both physical and psychological items

However, the QL of the survivors and their controls was explored by use of the SF-12 in the study by Moe and colleagues [28] and with the Rand Health Insurance study General Well-being measure by Tebbi and colleagues [22]. Neither of these studies found any statistical differences between the groups with respect to physical health and QL. However, Moe and colleagues [28] found that the somatisation score on the General Health Questionnaire with items closely related to fatigue demonstrated a significantly higher score for acute lymphoblastic leukaemia (ALL) survivors than for controls. Fatigue was also mentioned in the study by Wasserman and colleagues [38]. One of the physical residual effects, as reported by 5% of the Hodgkin's disease survivors, was easy fatigability. Nevertheless, the study by Zeltzer and colleagues [30] showed no difference between the POMS Fatigue subscale score of 552 ALL survivors and 394 sibling controls.

Apajasalo and colleagues [10] used the 15D (a 15-dimensional questionnaire) to examine the health-related QL of 168 survivors with a range of different malignancies and 129 controls. They found that the QL score of the survivors was significantly better than that of the controls; survivors reported better levels of vitality, distress, depression, discomfort, elimination and sleeping dimensions. There were no differences in QL between survivors with different malignancies, but it should be noted that the numbers in each diagnostic group were small.

Three studies attempted to measure physical functioning in bone tumour survivors. Two studies used a study-specific questionnaire [35,37], and the Karnofsky performance scale was used in the other study [36]. In all studies there is evidence that the bone tumour group had poorer physical functioning than their controls. These included specific difficulties with climbing stairs [35], and "general physical functioning" [36,37].

Psychological functioning

With respect to psychological functioning, we found that most studies focused on emotional aspects, using many different instruments. Most authors employed standardised measures with the availability of norms and comparison groups. In this section, emotional functioning is described in terms of overall emotional functioning, depression and anxiety, and self-esteem.

Overall emotional functioning
All investigations assessing the overall emotional or mental functioning of the survivors used standardised measures containing various dimensions of emotional well-being. In general, survivors seemed to be well adjusted. Most researchers found no difference in functioning between survivors and healthy peers and/or normative samples, based on the scores at the Rand Health Insurance Study General Well-Being measure [22], Multi-dimensional Personality Questionnaire Well-Being and Stress Reaction Scales [31], Profile of Mood States [8], SF-12 and General Health Questionnaire [28], Symptom Checklist-90-Revised (SCL-90-R) [13], Questionnaire on Subjective Well-Being [34], and Mental Health Inventory [12]. For the small percentage of survivors who did display one or

more clinical elevations on the SCL-90-R, three factors were identified which were associated with increased risk of maladjustment: older age at follow-up, greater number of relapses, and presence of severe functional impairment [13]. Survivors of bone tumors diagnosed in adolescence had more problems than survivors who became ill during childhood or early adulthood [34].

In two studies survivors appeared to be less well adjusted emotionally than their healthy peers or the general population. Lansky and colleagues [20], who used a structured interview to assess overall psychologic adjustment, reported a higher prevalence rate of episodes of treated depression, alcoholism and/or suicide attempts in survivors than in the general population. Both Gray and colleagues [8] and Zeltzer and colleagues [30] measured overall psychologic adaptation with the Profile of Mood States (POMS). While the first authors reported that 62 survivors with a range of diagnoses were similar to their 51 healthy age-matched peers, the 580 ALL survivors in the study by Zeltzer and colleagues had a greater negative mood, more tension, depression, anger and confusion than their 396 sex-matched siblings. The female survivors reported the highest mood disturbance. However, their scores were not as high as were found in a psychiatric sample. Finally, Elkin and colleagues [13] found that survivors' scores on the SCL-90-R subscales Anxiety, Psychoticism, Global Severity index, and Total Positive Symptoms were below normative values, suggesting that this group of survivors must be healthier than would be expected according to normative data.

Depression and anxiety
In some studies depression and anxiety were measured with a subscale of a standardised instrument measuring overall emotional adjustment. In most studies [8,12,13,23,28], no increased rates of depression and/or anxiety were reported. Zeltzer and colleagues [30], however, reported more depression among ALL survivors than among their siblings, and Lansky and colleagues [20] found that the prevalence of treated depression was higher in survivors than in the general population. Moreover, female survivors of ALL experienced more anxiety in stressful situations than the sex-appropriate norms, in contrast to males, who scored below the norms [31]. Felder-Puig and colleagues [34] used the scale "trait-anxiety" from the State-Trait-Anxiety Inventory in their study. No increased anxiety was found for the 26 survivors of bone tumors relative to the norm group.

In three studies, the Diagnostic and Statistical Manual of Mental Disorder (DSM) criteria were used to assess the frequency of affective disorders in survivors. Teta and colleagues [23] used the Schedule for Affective Disorder and Schizophrenia (SADS-L) and found that the prevalence of lifetime major depression in 450 survivors (with a variety of cancers) did not differ from that of their 587 sex-matched siblings. It was also similar to those reported in the literature for the general population. More recently, similar findings were reported by Mackie and colleagues [33], who found no increased rates of minor depression in 169 survivors of ALL or Wilms' tumors relative to 102 healthy age- and sex-matched controls. Finally, Wasserman and colleagues [38], who included a DSM psychiatric assessment in the interviews with 40 survivors of Hodgkin's disease, reported that the frequency of psychiatric diagnoses in the sample was basically no different from that found in community studies.

Self-esteem
In three studies assessing self-esteem with (a part of) a standardised instrument, no differences between survivors and control groups and/or normative groups were found. More specifically, the 60 bone tumour survivors in the study by Felder-Puig and colleagues [34] scored within normal ranges on the Frankfurt Self-Concept Scales, as did the 62 survivors on the Rosenberg Self-Esteem Scale [8]. The survivors in the latter study did not differ from their healthy peers. Finally, overall self-esteem of 48 survivors with a range of diagnoses was as high as that of their healthy siblings, as measured with the Oxford Psychologists Press adult self-esteem questionnaire [14].

Social functioning

Across studies, social functioning has been operationalised in a variety of ways, covering such issues as education, employment, insurance cover, living situation, marital status, and fertility, including reproductive capacity and family planning. Most investigations used (semi-) structured interviews with author-developed questionnaires.

Education
With respect to education, many research studies have demonstrated that survivors of childhood cancer, as a whole, did not differ much from controls or from the general population [12,14,16,19-21,28,34,36,38,39], although there were exceptions in certain subgroups of survivors. Kelaghan and colleagues [19] investigated the level of education in 2283 survivors and compared the results with those of 3261 sibling controls. The survivors of CNS tumors diagnosed before age 15 were significantly less likely than their controls to complete the eight grade of school. CNS tumour survivors who did complete secondary school were also less likely to enter college. The deficit was more severe in survivors who were treated with radiation therapy than those who underwent surgery alone. They also found that an early age at diagnosis was associated with a larger educational deficit than late age at diagnosis. Another study [16] reported that although 91% of the CNS tumour survivors had completed high school, only 10% had received a bachelor's or equivalent degree, as against 98% and 25%, respectively, in the non-CNS tumour group. Two studies evaluated the impact of treatment on scholastic performance in survivors of ALL [26,32]. Significantly more survivors than controls were placed in a special educational programme [26,32], or a learning disabled programme [26]. In the study by Kingma and colleagues [32] in ALL survivors with cranial radiotherapy (CRT) a significant difference in the level of secondary education was found for all survivor/sibling comparisons except in the case of survivors aged over 7 years at the time of diagnosis, when mean level of education no longer differed from that of their siblings. Younger age at diagnosis was also associated with referrals. The researchers found no effect of gender or irradiation dose on referral to special schools or on

level of secondary education. In contrast, Haupt and colleagues [26] reported that the risk associated with special education and learning-disabled programmes increased with increasing dose of CRT. Survivors treated with 24 Gy and those diagnosed before 6 years of age were less likely to enter college.

Finally, Evans & Radford [14] concluded from their study of 48 survivors with various tumours that there was no significant difference between survivors and siblings in qualifications at 16 years. However, survivors were significantly less likely to go on to higher education (16 years onwards) than their siblings. Many survivors (67%) felt that their education had suffered as a result of their disease. This percentage was higher than that found in the study by Dolgin and colleagues [12], in which 45% of the survivors reported that their illness had impacted on their educational achievement to a (very) great extent. In contrast, 77% of survivors in another study said that cancer had had no effect on their educational achievement [21].

Employment
The employment problems of cancer survivors have been of increasing interest during the last decades. Zeltzer and colleagues [30] studied 580 young adult survivors of ALL and found that significantly more survivors than sibling controls who had not enjoyed higher education were unemployed or were working less than half-time. This finding agrees partly with the study by Green and colleagues [15], who compared 227 former paediatric cancer patients with population norms. They found that the percentage of unemployed male survivors did not differ from population norms. The percentage of unemployed female suvivors, however, was slightly higher than that of the U.S. population in general. Other studies found that survivors and controls did not differ with respect to employment status [12,14,16,20,22,28,31,36] and that the majority of long-term survivors old enough to be in the work force were employed in a range of professional, clerical, and skilled labour positions [22,34]. Two studies looked specifically at survivors of bone tumors. Nicholson and colleagues [35] studied 111 survivors treated for Ewing's sarcoma and osteosarcoma and found that, in spite of a greater likelihood of having ever been disabled, their employment status did not differ from that of their siblings. Felder-Puig and colleagues [34], however, noted that many survivors treated for a bone tumour reported major difficulties in obtaining work.

In 1987, Mellette & Franco [41] reviewed the literature relating to employment of survivors of childhood cancer. They noted that, whereas in studies of a decade ago various forms of discrimination were reported, recent studies had been unable to document many of such problems. Nevertheless, Green and colleagues [15] found evidence of employment-related discrimination in 11% of 227 childhood cancer survivors who were treated between 1960 and 1985. Almost 30% of the male survivors were rejected for military service. However, these frequencies were lower than those reported by Teta and colleagues [23] and Wasserman and colleagues [38] in 1986 and 1987, respectively. Teta and colleagues reported in their study of 450 survivors and their 587 siblings that there was significantly more rejection of survivors (85%) than of their siblings (18%) by the military and other prospective employers (survivors 32%, siblings 21%). In the study by Wasserman and colleagues of 40 survivors of childhood and adolescent Hodgkin's disease, 20% reported that they had experienced job discrimination. In a recent study by Dolgin and colleagues [12], 46% of the Israeli survivors reported that their illness had impacted on their employment histories "to a great extent" or "to a very great extent". Forty-five percent of the survivors had been rejected from a workplace, compared with 19% of the controls. Approximately half of these survivors felt that their workplace rejection was due to their cancer history. They also found that 55% of the survivors had difficulty being accepted into the military service. Rejection for the military has also been reported in another investigation [16].

Six studies have assessed the level of income. Dolgin and colleagues [12] and Hays [16] found that survivors reported less annual income than the controls, however, in the latter study this difference was not significant. Interestingly, the survivors in the study by Tebbi and colleagues [22] reported a higher mean income than controls. The other studies found no differences [14,34,35].

Insurance
Obtaining adequate health and life insurance has been a recurring problem for survivors of cancer. Although the differences were not significant, male and female survivors reported they were turned down for life and health insurance more frequently than their siblings [23]. A report of insurance problems among 100 survivors who were treated during the years 1945-1975 showed that 24% had difficulty in securing health insurance and 15% had no health insurance at the time of the survey, versus 0% and 7%, respectively, in these categories among controls [17]. Difficulty in obtaining life insurance was noted by 44% of survivors and by only 2% of matched controls. Tebbi and colleagues [22] found that many survivors had difficulty in obtaining health, life, or disability insurance. Green and colleagues [15] found that the percentages of survivors who had life insurance and company-offered health insurance were lower than those reported for the general U.S. population. Twenty-four percent of those with life insurance had had difficulty in obtaining it. Although a small percentage (7%) of survivors in the study by Hays and colleagues [16] had been denied employment-related health insurance at some time and another 8% had at some time had health insurance cover that excluded cancer, most survivors were covered by health insurance policies without cancer-related restrictions. There were no differences from the controls. Evidence of both past and current discrimination in obtaining affordable life insurance on the basis of a cancer history was clearly recognisable. However, the majority of survivors who desired life insurance were insured and at standard rates. Novakovic and colleagues [36] found no difference in health care insurance status, but more problems in getting job-related health insurance. Finally, Jacobson Vann and colleagues [18] assessed the effects of having a cancer diagnosis on the subsequent acquisition of health insurance cover for young adults diagnosed as children in North Carolina. They found that survivors were turned down for health insurance cover more often than their siblings, this was

due, according to the survivors, to their cancer history and related medical history. Survivors were also more likely than their siblings to have health insurance policies with clauses excluding cover for pre-existing health conditions. When participants were asked whether they had had problems in obtaining health insurance coverage, 24% survivors answered "yes", as apposed to 2% of the responding siblings. Furthermore, survivors were 4.3 times as likely to be covered by their parents' health insurance policies.

Living situation, marital status and family
Only two investigations have specifically addressed the living situation of young adult survivors. In a pilot study of 39 survivors Lansky and colleagues [20] found that survivors did not significantly differ from the sibling group on living arrangements (with parents versus other); however, the survivors left home at a slightly older age (21 versus 19 years). The survivors in Felder-Puig's study [34] also seemed to stay at home longer after reaching adulthood than controls of a similar age.

Two studies have focused solely on marriage issues among childhood cancer survivors, and several studies of the late effects on cancer treatment have included data on marital status as an indicator of social competence. The largest and most comprehensive study of marriage, which compared 10425 survivors with a broad range of diagnoses with U.S. population norms, was published by Rauck and colleagues in 1999 [25]. They found that the percentage of survivors who had ever been married was lower than that in the general U.S. population within similar age groups. In particular, compared with their age-matched counterparts in the general population, women and whites were less likely to have married, whereas black survivors were more likely to have married. Comparison of childhood tumour types showed that survivors with a diagnosis of CNS tumors, particularly males, were less likely to have married than those with other diagnoses or the general population. In the second largest study of marriage, which compared 2170 survivors with sibling controls, Byrne and colleagues [11] also found that, as a group, survivors were less likely to be married and that the differences were greatest among male survivors of CNS tumors. Similar findings were reported in some smaller studies. Zevon and colleagues [31] found a decreased frequency of marriage for both men and women relative to the U.S. population in a group of 46 survivors with ALL. These conclusions were supported in a study of 227 survivors, including few with a diagnosis of CNS tumour by Green and colleagues [15] and in two other studies with survivors of bone cancer [34,36]. Green and colleagues [15] also found that marital status was not affected by age at diagnosis, gender, history of disease recurrence and diagnosis. Teeter and colleagues [24] reported data collected by the University of Kansas on marital status among 263 survivors and 369 controls. Twenty-five percent of the survivors and 16 percent of the controls had never married. Makipernaa [39] studied survivors diagnosed with solid tumors (excluding CNS tumours) and found that fewer of the women and as many of the men were married as in the general population. Finally, in a study of 95 survivors, Meadows and colleagues [21] found that survivors were less likely to be married than members of the sibling control group. However, the authors stated that this was probably a biased comparison, because the siblings as a group were older than the survivors.

In contrast, other studies have suggested that there are no significant differences among survivor/control comparisons with respect to marital status. For example, Nicholson and colleagues [40] found no marriage relative to controls in a population of 111 survivors of bone cancer. Hays and colleagues [16] reported marriage statistics from two centres, which showed no difference between survivors with a variety of diagnoses and the general U.S. population when CNS tumors were excluded. Wasserman and colleagues [38] studied 40 survivors of childhood and adolescent Hodgkin's disease and also found that the overall proportions of marriage in the survivors were not different from the general population statistics. Four other studies yielded similar results [12,14,20,30].

Some studies provide data on specific reasons for not marrying. All the participants in the study by Teeter and colleagues [24], were asked whether they had refrained from marrying for medical or health reasons. Twenty-one survivor (31%) and one control (2%) said that they had not married for health reasons. Green and colleagues [15] found that among the survivors who had never married or lived as married (n=96), almost 16% reported that their history of childhood cancer had influenced their decisions on marriage. In Makipernaa's study [39] 5 survivors reported that it was expressly the cancer treatment that had made them decide to remain single. One woman emphasised that knowing she had had a hysterectomy had prevented her marriage. Four others felt that the cancer and its treatment had so impaired their appearance that it hampered their personal contacts. Most single survivors in the study by Meadows and colleagues [21] indicated that having had cancer had no impact on their desires or opportunities for marriage. However, 21% said that having had cancer sometimes affected their ability to meet others, and 38% reported that their history of cancer sometimes scares others.

There was no significant difference in the overall frequency of separation or divorce in the study by Green and colleagues [15]. However, a more detailed analysis of the separation and divorce data revealed that the percentage of divorced women aged 35-44 was significantly greater relative to that in the normative group. Zevon and colleagues [31] also reported an elevated frequency of divorce in women compared with the rates for the general population. In contrast, separate analysis of the men in the study of Wasserman and colleagues [38] showed a significantly higher rate of divorce than in age- and race-specific statistics. Survivors of bone tumors were also found to be more likely to have divorced in the study by Novakovic and colleagues [36]. One study found that, in general, the proportion of survivors who were divorced or separated was lower than that of the U.S. population [25]. Men, however, were more likely to have divorced or separated than their age-matched counterparts in the general population, and women less likely. Survivors with the diagnosis of CNS tumour were also more often divorced or separated than those with other diagnoses or the general population. The latter finding was confirmed by Hays and colleagues [16] who found that in the CNS group 23% survivors had been divorced, versus 8% in the non-CNS tumour category. Byrne and colleagues [11] also reported that first marriages of male survivors of CNS tumors who were diagnosed before age 10 years were three times as likely to end than those of controls. They also found that male survivors of retinoblastoma had higher divorce rates than male controls.

The effect of a history of childhood cancer on divorce was addressed in one study [15]. For those survivors who were separated or divorced, 20% (n=5) reported that their history of childhood cancer had been a contributing factor to the dissolution of their relationships.

The issue of fertility has been investigated by several investigators. Nicholson and colleagues [35] found that although deficits in crude fertility rates were significant when all former bone cancer survivors were compared against all controls, these differences were non-significant after controlling for sex. According to Moe and colleagues [28] men once diagnosed with ALL had significantly fewer offspring than the men in the control group, whereas the women in the ALL group had slightly more children than their controls. Three other studies reported that the percentage of survivors with children was lower than the percentage of controls [16,24,36]. Among the survivors who had ever been married or lived as married in the study by Green and colleagues [15], 10% indicated that their history of childhood cancer influenced their decision to limit the number of children they had to a moderate or greater degree. For an additional 10%, their medical history was a factor that contributed to their decision to have no children. Worries about reproductive capacity were reported in three studies. Gray and colleagues [8] found that survivors were more likely than a matched control group of peers to report worrying about being able to conceive a child. When Wasserman and colleagues [38] conducted open-ended interviews in 40 adult survivors of Hodgkin's disease, they found that female survivors often reported concerns about fertility, whereas male survivors did so much less often. Forty-six percent of the female ALL survivors and 29% of the male ALL survivors in the study by Zevon and colleagues [31] reported being concerned about possible future health problems their children might experience as a result of their cancer history.

Sexual functioning

So far, not many studies provide data on sexual functioning. Veenstra and colleagues [37] assessed body image and sexual functioning in 33 bone tumour survivors with a rotation plasty. Almost half of the survivors felt slightly to very limited in initiating intimate relationships as a result of the rotation plasty. While 19 survivors reported that they did not feel physically unattractive as a result of the rotation plasty during the week prior to the assessment, 10 reported feeling a little unattractive and 4 reported feeling quite a bit to very unattractive. Of the survivors who were sexually active (n=21), 10 survivors reported that they were limited in their sexual activities to a small (n=8) or moderate (n=2) degree as a result of the surgery.

Puukko and colleagues [29] investigated possible changes in sexual identity, sexual attitudes, and sexual behavior of 30 female survivors diagnosed with acute leukaemia as compared with healthy age-matched controls. They found that survivors did not differ from controls with respect to the following aspects of sexual behavior: age at which dating began, onset and frequency of sexual intercourse, and opinions on sexual behavior. They also found that there were significant differences in behavior: survivors were less likely to have experienced sexual intercourse, less likely to have initiated intercourse, less likely to masturbate and less likely to have talked with friends about sexual topics. With regard to inner sexuality, survivors also differed from controls. Their images of sexuality were more restrictive, and their attitudes, especially those concerning sexual pleasure, were more negative than those of the controls. Finally, sexual identity among the survivors was less often feminine and more often infantile than among controls.

Factors related to survivors' functioning

Fortunately, not all young adult survivors of childhood cancer seem to suffer from the late sequelae of their disease and/or treatment. So it is very important to identify factors that predict good QL and to trace risk factors. In most studies factors related to survivors' function have been discussed to some extent. Predictors can be divided into demographics, and illness- and treatment related factors.

Demographics
In several studies gender has been investigated in relation to survivors' functioning. Especially female survivors seemed to be at risk for psychological problems. According to Zevon and colleagues [31] female ALL survivors had an increased tendency to experience anxiety in stressful situations, and in the study by Zeltzer and colleagues [30] female ALL survivors reported the greatest total mood disturbance. With respect to marriage, the percentage of married female survivors was lower [15,25], but according to Rauck and colleagues [25] female survivors were less likely to divorce/separate. However, Green and colleagues [15] found that a subgroup of female survivors (aged 35-44) had a significantly higher frequency of divorce than age-specific group norms. Male gender was positively related to employment [15]. In contrast with these findings, Apajasalo and colleagues [10], Elkin and colleagues [13] and Kingma and colleagues [32] reported that gender was not associated with survivors' functioning respectively with QL, maladjustment according to the SCL-90-R, and educational status.
In five studies, age at study has been analysed in relation to outcome. Age was found to be negatively related to psychological functioning. Younger survivors reported a better QL [10], and older survivors scored higher on the Symptom Checklist [13]. Older survivors were also more likely never to have married than younger survivors [30], but they were more likely to be employed [15]. The sexual identity of the survivors seemed not to be associated with age according to Puukko and colleagues [29].

Two studies reported results about minority survivors. Minority survivors of ALL showed the highest mood disturbance [30]. Black survivors were generally found to be more likely to have married, but also more likely to have divorced/separated once married than the general US population [25].

Illness and treatment related factors
Age at diagnosis is one of the factors that has been most frequently investigated in relation to survivors' functioning. Survivors diagnosed at a younger age were at higher risk for poor educational performance [19,26,32]. Felder-Puig [34] concluded that survivors of bone tumors diagnosed in adolescence had more problems (especially less social well-being) than those diagnosed in childhood or early adulthood. In contrast, in seven other studies age at diagnosis appeared to be not associated with outcome, and/or not to be related to emotional functioning [8], depression [23], maladjustment in terms of the SCL-90 [13], poor close relationships [33], sexual identity [29], educational level [21], or marriage, divorce, employment and insurance [15].

With respect to the diagnosis (type of cancer), especially CNS tumors versus other diagnoses is the comparison that has been most intensively investigated. It was found that survivors of CNS tumors were more seriously affected. Their educational level was lower [8,16,19], and they were less likely to be married [11,16]. Moreover, they were more likely to have divorced and their rates of parenthood were also lower [16]. Elkin and colleagues [13], who studied survivors with a range of diagnoses, found no relation between type of cancer and QL. Similar findings were reported by Apajasalo and colleagues [10], who excluded survivors with CNS tumors.

With respect to type of treatment, radiation therapy appeared to be a risk factor. First, survivors who were treated with cranial irradiation showed less well-being than the other ALL survivors [31]. Second, treatment with radiation therapy versus surgery alone , and a higher dose of cranial irradiation [26] seemed to be risk factors for poor educational performance. However, among the ALL survivors in the study by Kingma and colleagues [32], the cranial irradiation was not associated with educational level.

In a sample of survivors with a range of diagnoses (except CNS tumours) survivors of bone marrow transplantation has a slightly lower QL than the other survivors [10]. In three other samples with a variety of diagnoses no association was found between the type of treatment and outcome: emotional functioning according to the SCL-90 [13], depression [23], educational level [21]. Moreover, Puukko and colleagues [29] concluded that the sexual identity of ALL survivors was not predicted by the type of treatment and Novakovic and colleagues [36] found that the treatment protocol of bone tumour survivors was not related to marriage and having children.

According to Mackie and colleagues [33] longer duration of treatment in survivors of ALL and Wilms' tumour was related to poor close relationships. In the same study this was also found in survivors whose illness was more recent. In three studies in which time since diagnosis or time since end of treatment was investigated no association with outcome was found [8,10,29].

Only three studies looked at the effect of recurrence of the disease. While Elkin and colleagues [13] concluded that disease relapse was a risk factor for emotional maladjustment, the opposite was found in the study by Gray and colleagues [8]. Green and colleagues [15] also found no evidence that recurrence of cancer was related to survivors' functioning, specifically it was not related to marriage, divorce, employment and insurance.

Medical and functional late effects were investigated in two samples of survivors of bone tumors. Nicholson and colleagues [40] reported that amputation status was not associated with health perception and Felder-Puig and colleagues [34] found no correlation between emotional functioning with physical or functional sequelae. In line with these results, disability, which was reported by survivors with different tumours, was not related to emotional functioning [8]. In contrast, Elkin and colleagues [13] concluded that severe functional impairment was a risk factor for maladjustment, while cosmetic status was not.

Conclusion and future directions

The purpose of this review was to give an overview of the research about QL in young adult survivors of childhood cancer populations during the last two decades. This review identified a wide variety of studies. Studies are characterised by a high degree of heterogeneity with respect to: the patients samples investigated (e.g. survivors with different cancers who had undergone a variety of treatments), the comparison groups selected, the QL dimensions assessed and the instruments employed. Additionally, age at time of evaluation, age at diagnosis, and time elapsed since completion of therapy varied widely. Moreover, the majority of the studies reviewed suffered from at least one of the following methodological weaknesses: small samples, nonstandardised, study-specific instruments, and cross-sectional rather than prospective designs. Given all of these differences between studies, perhaps it is not surprising that outcomes of studies differ and that the QL reported by survivors also varies, making it impossible at this time to come to firm conclusions about the magnitude and nature of long-term consequences for childhood cancer survivors.

However, despite the heterogeneity in study procedures and the methodological shortcomings, some clear trends emerge from this review. Although some inconsistent data have been reported across studies, the results suggest the following.

Physical functioning
1) The majority of survivors reported they were in apparently good health, with the exception of bone tumour survivors, who were more likely to perceive their health as fair or poor. Bone tumour survivors also had poorer physical

functioning than their controls. Difficulties in climbing stairs and poor "general physical functioning" were reported.
2) Some studies mentioned fatigue as a residual effect of treatment.

Psychological functioning
1) Most long-term survivors functioned well psychologically and did not have significant more emotional problems than controls. The subgroup of survivors who reported problems mentioned depression, mood disturbances, tension, anger, confusion and anxiety. Female gender, older age at follow-up, greater number of relapses, presence of severe functional impairment, cranial irradiation and minority survivors were associated with an increased risk for emotional problems in some studies.

Social functioning
1) Survivors of CNS tumours and survivors of ALL seemed to be at risk for educational deficits. Cranial irradiation and an early age at diagnosis was associated with educational deficits. Many survivors reported that their education had suffered as a result of their disease.
2) The majority of survivors old enough to be in the workforce were employed. Although in almost all research survivors did not differ from controls with respect to employment status, some survivors experienced some form of job discrimination and difficulties in obtaining work. Problems in obtaining health and life insurance were also reported.
3) Survivors seem to stay at home longer after reaching adulthood and leave home at an older age than their controls.
4) There is a lower prevalence of marriage among survivors, particularly in male survivors with a diagnosis of CNS tumors. The survivors reported that the history of childhood cancer, the consequences of treatment and problems with health as specific reasons for not marrying.
5) The percentage of survivors with children seems lower. The survivors indicated that the medical history is a factor that contributes to the decision to have no children. Many survivors reported worrying about their reproductive capacity and/or about possible future health problems their children might experience as a result of their cancer history.

Childhood cancer was almost always a fatal disease in the not-too-distant past. Over the last decades significant treatment advances have been made, and long-term survival is now a reality. With the increasing number of long-term childhood cancer survivors, the need to assess their QL is becoming more important and meaningful. This article has summarised what is known about the long-term effects of disease and treatment on the QL of survivors. Where do we go from here?

It is evident that additional research is needed. Although the low incidence of childhood cancer, the variety of diseases and treatments and the wide range in ages pose methodological problems in QL assessment, we need well-designed studies. Since not many institutions have a sufficient number of patients to control for the numerous patient-specific and therapy-specific variables involved, multi-institutional collaboration is recommended. At the least, account must be taken of the age of the child at diagnosis and treatment, the length of time since completion of therapy, and the differences in severity of the cancer and its treatment, and thus the treatment era. The QL dimensions of interest, and therefore the outcome measures of the study, must be clearly defined. This will enhance the possibility of comparing international studies and to conducting systematic reviews. Researchers should attempt to use prospective study designs with sufficiently large sample sizes, choose instruments appropriate to their goals and establish the methodological properties of the instruments they use in keeping with that goal. However, in this still-evolving area of research, it is wise for investigators to include an opportunity for survivors to report additional concerns not covered in standardised QL scales wherever possible. Naturally, one or another is dependent on the question of whether the main objective is measurement of differences between patients at one particular point in time or longitudinal change within patients over time.

As Gotay & Muraoka [42] stated in their review on QL in adult-onset cancer survivors, there is a need to understand the long-term impact of different treatment on QL. It is important to document how varying therapeutic modalities can give rise to different long-term effects. Such information can establish whether there are any residual effects of one treatment but not another and whether there are treatment-related decrements in QL that vary in the short term and long term. Further, little is known about the impact of persistent effects of cancer treatment on survivors' QL. Survivors may learn to live with and adjust to their possible limitations, they may continue to experience problems to the same degree as during short-term survival, or their tolerance of disability may decline with the passage of time (i.e., an enhanced QL, an unchanged QL, or a worsened QL, respectively) [42]. It is also important to identify the subgroups of survivors who have problems rather than evaluating only differences between survivors as a whole and their controls.

Many of the studies reported to date are based on North-American samples; this seems to be an area of research in which North American researchers have taken a lead. However, there are many cultural differences between the United States and European countries, in addition to dissimilarities in their health care systems, particularly with respect to health care insurance. No studies were found for this review from anywhere outside the United States and Europe, and this raises questions about the functioning of childhood cancer survivors in other countries. The increasing cultural/ethnic diversity of people within the borders of all countries and the growing communication network around the globe underscore the relevance of cross-cultural comparisons [43]. It is known that there are many differences in adjustment to cancer across cultures. The cultural distinctions between and within national boundaries provide an unique opportunity to examine differences in the meaning of survivorship, as well as values and behaviours, in different groups.

Future research efforts should also be directed at the intermediate factors of QL that have received relatively little attention in previous studies, including the role of coping and adaptation, social relationships and family variables. Considering that many survivors are functioning reasonably well and that not much difference is found between results in survivors and their peers, it would be interesting and advisable to investigate the role of denial and response-shift. It would also be interesting to know if survivors meet developmental tasks in growing up. More insight is therefore needed into the relation between the survivors' course of life and their functioning in later life. The need for future studies applies for other aspects as well, such as posttraumatic stress, body image and spiritual dimensions. Although these concepts are investigated in younger survivors, we did not find any studies in which these topics had been measured with standardised questionnaires and compared with norms in young adults. It is also remarkable that no studies about cognitive functioning in (young) adult survivors of childhood cancer were found, in contrast with the large number of studies done in children. As Kingma [27] has mentioned, it is not yet known what may happen to maturing brains long after exposure to CRT and/or chemotherapy in childhood. Furthermore, in the adult cancer literature it is suggested that more research is needed because neuropsychologic symptoms, particularly problems with memory and concentration, are frequently reported by cancer patients treated with chemotherapy, even years after completion of treatment [44].

As we learn more about the challenges associated with long-term childhood cancer survival, interventions will be needed to address the problems identified. It is possible that some problems can be prevented and others remediated if appropriate care is provided. However, it is critical to determine the kind of support desired by long-term survivors and to identify who is most in need of and likely to benefit from such interventions [45,46]. Therefore, it is critical to ask survivors what they need and what they want, for example by means of focus groups. Interventions to reduce psychological morbidity or improving QL, such as patient education, coping skills management, and support groups deserve continued attention. Studies are needed to identify the extent to which these interventions improve QL.

Acknowledgements
The authors thank Dr. Cor van den Bos for his valuable advice.

References

1. Stiller CA. Population based survival rates for childhood cancer in Britain, 1980-1991. *BMJ* 1994, **309**, 1612-1616.
2. Hawkins MM, Stevens MC. The long-term survivors. *Br Med Bull* 1996, **52**, 898-923.
3. Hobbie W, Ruccione K, Moore IK, Truesdell S. Late effects in long-term survivors. In: Foley GV, Fochtman D, Mooney KH, eds. Nursing Care of the Child with Cancer. Orlando, Florida: W.B. Saunders Company 1993, 466-496.
4. Schwartz CL, Hobbie WL, Constine LS, Ruccione KS, eds. Survivors of Childhood Cancer: Assessment and Management. St Louis, Missouri: Mosby-Year Book, Inc 1994.
5. World Health Organization DoMH. WHO-QOL Study protocol: The development of the World Health Organization quality of life assessment instrument. Geneva, Switzerland 1993
6. Leigh SA, Stovall EL. Cancer Survivorship. Quality of Life. In: King CR, Hinds PS, eds. Quality of Life. From Nursing and Patient Perspectives. Sudbury, Massachusetts: Jones and Bartlett Publishers 1998, 287-300.
7. Eiser C, Hill JJ, Vance YH. Examining the psychological consequences of surviving childhood cancer: systematic review as a research method in pediatric psychology. *J Pediatr Psychol* 2000, **25**, 449-460.
8. Gray RE, Doan BD, Shermer P, FitzGerald AV, Berry MP, Jenkin D, Doherty MA. Psychologic adaptation of survivors of childhood cancer. *Cancer* 1992, **70**, 2713-2721.
9. Gray RE, Doan BD, Shermer P, FitzGerald AV, Berry MP, Jenkin D, Doherty MA. Surviving childhood cancer: a descriptive approach to understanding the impact of life-threatening illness. *Psychooncology* 1992, **1**, 235-245.
10. Apajasalo M, Sintonen H, Siimes MA, Holmberg C, Boyd H, Makela A, Rautonen J. Health-related quality of life of adults surviving malignancies in childhood. *Eur J Cancer* 1996, **32A**, 1354-1358.
11. Byrne J, Fears TR, Steinhorn SC, Mulvihill JJ, Connelly RR, Austin DF, Holmes GF, Holmes FF, Latourette HB, Teta J, Strong LC, Myers MH. Marriage and divorce after childhood and adolescent cancer. *JAMA* 1989, **262**, 2693-2699.
12. Dolgin MJ, Somer E, Buchvald E, Zaizov R . Quality of life in adult survivors of childhood cancer . *Soc Work Health Care* 1999, **28**, 31-43.
13. Elkin TD, Phipps S, Mulhern RK, Fairclough D. Psychological functioning of adolescent and young adult survivors of pediatric malignancy. *Med Pediatr Oncol* 1997, **29**, 582-588.
14. Evans SE, Radford M. Current lifestyle of young adults treated for cancer in childhood. *Arch Dis Child* 1995, **72**, 423-426.
15. Green DM, Zevon MA, Hall B. Achievement of life goals by adult survivors of modern treatment for childhood cancer. *Cancer* 1991, **67**, 206-213.
16. Hays DM, Landsverk J, Sallan SE, Hewett KD, Patenaude AF, Schoonover D, Zilber SL, Ruccione K, Siegel SE. Educational, occupational, and insurance status of childhood cancer survivors in their fourth and fifth decades of life. *J Clin Oncol* 1992, **10**, 1397-1406.
17. Holmes GE, Baker A, Hassanein RS, Bovee EC, Mulvihill JJ, Myers MH, Holmes FF. The availability of insurance to long-time survivors of childhood cancer. *Cancer* 1986, **57**, 190-193.
18. Jacobson Vann JC, Biddle AK, Daeschner CW, Chaffee S, Gold SH. Health insurance access to young adult survivors of childhood cancer in N. Carolina. *Med Pediatr Oncol* 1995, **25**, 389-395.
19. Kelaghan J, Myers MH, Mulvihill JJ, Byrne J, Connelly RR, Austin DF, Strong LC, Wister Meigs J, Latourette HB, Holmes GF, Holmes FF. Educational achievement of long-term survivors of childhood and adolescent cancer. *Med Pediatr Oncol* 1988, **16**, 320-326.
20. Lansky SB, List MA, Ritter-Sterr C. Psychosocial consequences of cure. *Cancer* 1986, **58**, 529-533.
21. Meadows AT, McKee L, Kazak AE. Psychosocial status of young adult survivors of childhood cancer: a survey. *Med Pediatr Oncol* 1989, **17**, 466-470.
22. Tebbi CK, Bromberg C, Piedmonte M. Long-term vocational adjustment of cancer patients diagnosed during adolescence. *Cancer* 1989, **63**, 213-218.
23. Teta MJ, Del Po MC, Kasl SV, Meigs JW, Myers MH, Mulvihill JJ. Psychosocial consequences of childhood and adolescent cancer survival. *J Chron Dis* 1986, **39**, 751-759.
24. Teeter MA, Holmes GE, Holmes FF, Baker AB. Decisions about marriage and family among survivors of childhood cancer. *J Psychosoc Oncol* 1987, **5**, 59-68.

25. Rauck AM, Green DM, Yasui Y, Mertens A, Robison LL. Marriage in the survivors of childhood cancer: a preliminary description from the Childhood Cancer Survivor Study. *Med Ped Oncol* 1999, **33**, 60-63.
26. Haupt R, Fears TR, Robison LL, Mills JL, Nicholson HS, Zeltzer LK, Meadows AT, Byrne J. Educational attainment in long-term survivors of childhood acute lymphoblastic leukemia. *JAMA* 1994, **272**, 1427-1432.
27. Kingma A. Neuropsychological late effects of leukemia treatment in children. Thesis. Groningen Rijksuniversiteit, The Netherlands 2001.
28. Moe PJ, Holen A, Glomstein A, Madsen B, Hellebostad M, Stokland T, Wefring KW, Steen-Johnsen J, Nielsen B, Howlid H, Borsting S, Hapnes C. Long-term survival and quality of life in patients treated with a national all protocol 15-20 years earlier: IDM/HDM and late effects? *Pediatr Hematol Oncol* 1997, **14**, 513-524.
29. Puukko LR, Hirvonen E, Aalberg V, Hovi L, Rautonen J, Siimes MA. Sexuality of young women surviving leukaemia. *Arch Dis Child* 1997, **76**, 197-202.
30. Zeltzer LK, Chen E, Weiss R, Guo MD, Robison LL, Meadows AT, Mills JL, Nicholson HS, Byrne J. Comparison of psychologic outcome in adult survivors of childhood acute lymphoblastic leukemia versus sibling controls: a cooperative Children's Cancer Group and National Institutes of Health study. *J Clin Oncol* 1997, **15**, 547-556.
31. Zevon MA, Neubauer NA, Green DM. Adjustment and vocational satisfaction of patients treated during childhood or adolescence for acute lymphoblastic leukemia. *Am J Pediatr Hematol Oncol* 1990, **12**, 454-461.
32. Kingma A, Rammeloo LA, Der Does-van den Berg A, Rekers-Mombarg L, Postma A. Academic career after treatment for acute lymphoblastic leukaemia. *Arch Dis Child* 2000, **82**, 353-357.
33. Mackie E, Hill J, Kondryn H, McNally R. Adult psychosocial outcomes in long-term survivors of acute lymphoblastic leukaemia and Wilms' tumour: a controlled study. *Lancet* 2000, **355**, 1310-1314.
34. Felder-Puig R, Formann AK, Mildner A, Bretschneider W, Bucher B, Windhager R, Zoubek A, Puig S, Topf R. Quality of life and psychosocial adjustment of young patients after treatment of bone cancer. *Cancer* 1998, **83**, 69-75.
35. Nicholson HS, Mulvihill JJ, Byrne J. Late effects of therapy in adult survivors of osteosarcoma and Ewing's sarcoma. *Med Pediatr Oncol* 1992, **20**, 6-12.
36. Novakovic B, Fears TR, Horowitz ME, Tucker MA, Wexler, LH. Late effects of therapy in survivors of Ewing's sarcoma family tumors. *J Pediatr Hematol Oncol* 1997, **19**, 220-225.
37. Veenstra KM, Sprangers MA, van der Eyken JW, Taminiau AH. Quality of life in survivors with a Van Ness-Borggreve rotationplasty after bone tumour resection. *J Surg Oncol* 2000, **73**, 192-197.
38. Wasserman AL, Thompson EI, Wilimas JA, Fairclough DL. The psychological status of survivors of childhood/adolescent Hodgkin's disease. *Am J Dis Child* 1987, **141**, 626-631.
39. Makipernaa A. Long-term quality of life and psychosocial coping after treatment of solid tumours in childhood. A population-based study of 94 patients 11-28 years after their diagnosis. *Acta Paediatr Scand* 1989, **78**, 728-735.
40. Nicholson HS, Byrne J. Fertility and pregnancy after treatment for cancer during childhood or adolescence. *Cancer* 1993, **71**, 3392-3399.
41. Mellette SJ, Franco PC. Psychosocial barriers to employment of the cancer survivor. *J Psychosoc Oncol* 1987, **5**, 97-115.
42. Gotay CC, Muraoka MY. Quality of life in long-term survivors of adult-onset cancers. *J Natl Cancer Inst* 1998, **90**, 656-667.
43. Padilla GV, Kagawa-Singer M. Quality of life and culture. In: King CR, Hinds PS, eds. Quality of life from nursing and patient perspectives: theory, research, practice. Sudbury, Massachusetts: Jones and Bartlett Publishers 1998, 74-92.
44. Schagen SB, van Dam FSAM, Muller MJ, Boogerd W, Lindeboom J, Bruning PF. Cognitive deficits after postoperative adjuvant chemotherapy for breast carcinoma. *Cancer* 1999, **85**, 640-650.
45. Rose JH. Social support and cancer: adult patients' desire for support from family, friends, and health professionals. *Am J Community Psychol* 1990, **18**, 439-464.
46. Worden JW, Weisman AD. Do cancer patients really want counseling? *Gen Hosp Psychiatry* 1980, **2**, 100-103.

Chapter 4

"I DON'T HAVE ANY ENERGY"
The experience of fatigue in young adult survivors of childhood cancer

Abstract

<u>Purpose</u>: Although it is speculated that fatigue occurs equally in adults as well as in children and adolescents with cancer, little research exists to substantiate this view. Evidence that fatigue continues after treatment is limited both in the adult and paediatric oncology literature. Due to the current lack of knowledge, more information on the phenomenology of fatigue of childhood cancer survivors is desirable. Therefore a study was conducted to explore the concept of fatigue from a survivors perspective.

<u>Patients and methods</u>: A semi-structured interview was conducted with a purposeful sample of 35 long-term survivors of childhood cancer who reported feeling extremely fatigued. The topics which were covered during the interview included the nature, onset and pattern of fatigue, sleep rest pattern, what helps with fatigue and what does not help, and the impact of fatigue on their daily life.

<u>Results</u>: Most survivors who were diagnosed with cancer in their adolescence identified fatigue as a significant side-effect of the treatment. The majority of survivors who were toddler or preschooler at the time of cancer treatment, mentioned that as far as they could recall, they had suffered from fatigue their entire life. The course of fatigue during the day differed among the survivors, although the majority reported to be fatigued when waking up in the morning. None of the survivors reported sleep problems. Many survivors slept 9 hours or more. Fatigue was defined by all respondents as having a negative impact on their daily lives.

<u>Conclusion</u>: Findings revealed that fatigue is a serious problem for some young adult survivors of childhood cancer and affects many aspects of quality of life.

Introduction

Since the 1970's, treatment results in children with cancer have significantly improved. Nowadays, the cure rate is about 70%, and in Western countries 1 in every 1000 young adults is a person who has been cured [1].

This success, however, has an unpleasant reverse. In the past decade it has become evident that former anti-cancer treatment, especially radiotherapy and some chemotherapeutic drugs, can cause harmful side-effects in many survivors. Often those side-effects become evident many years after the treatment [2,3]. These late sequelae may affect organs, for example damage the heart, lungs, kidneys, bones, hormone systems or reproductive system. Neuropsychological studies which focus on the long-term effects of cranial exposure to radiation on the cognitive functioning of survivors often document some impairment [4]. Psychosocial problems may develop as a consequence of these disabilities. Besides these late sequelae, survivors of childhood cancer are at risk of a second malignancy [5].

Therefore, it is important to carefully monitor the sequelae in survivors of childhood cancer. In Amsterdam, survivors are followed annually according to protocols and with special attention to the side-effects. Since 1996, these persons have been examined by a specialized follow-up team which consists of a paediatric oncologist, an internist-oncologist and a research nurse. In the past 3 years about 450 survivors were seen by the team. As expected, we saw the whole range of side- effects, varying from mild to very severe, even life-threatening conditions. Unexpectedly, however, a subgroup of these persons complained about extreme fatigue. The complaints were expressed with such uniformity, that we suspected that we were dealing with a pathological entity related to the former cancer or its treatment. This stimulated us to explore the experience of fatigue in this group.

Researchers agree that fatigue related to cancer is a subjective, multidimensional sensation which is similar to pain, and which can be best measured by self-report. Fatigue can be defined as the perception of an unusual or debilitating sense of wholebody tiredness, different from the usual sense of tiredness experienced by healthy individuals [6]. Furthermore, fatigue is multifactorial. Piper [7] (p 220) defined fatigue, from a nursing perspective, as "a subjective feeling of tiredness that is influenced by circadian rythm and can vary in unpleasantness, duration, and intensity".

The aetiology of fatigue in cancer patients is complex. Factors, which are generally associated with fatigue in cancer patients are generally classified into 3 categories: physiological, psychological, and situational factors [8]. Physiological factors include anaemia, chronic pain, nutritional status and various biochemical changes secondary to disease and treatment [8]. In general, fatigue is attributed to treatments such as surgery [9] and bone-marrow transplantation [10]. Fatigue is reported by 60 to 96% of the patients who receive chemotherapy [11-14]. The majority of the patients who receive radiation therapy, experience fatigue that worsens over time, regardless of the field of radiation [11,14-17] . Furthermore, fatigue is a dose-limiting toxicity for many biological response modifiers (interferons or interleukins), because patients often become so exhausted that they opt to discontinue treatment [18,19].

Psychological factors, such as anxiety or depression can lead to fatigue [20,21]. The relationship between depression and fatigue in cancer patients is known to be complex. Fatigue may be the result of a depressed mood [22]. However, the person who continuously perceives his or her energy as insufficient may become depressed. To complicate matters, in cancer depression and fatigue may co-occur without having a causal relationship, because they can both originate from the same pathology [23]. Immobility, crisis or problems with relationships are the most common situational factors that contribute to fatigue in cancer patients [8]. Results of a recent tripart survey on the perceptions of cancer-related fatigue in 419 cancer patients, 200 primary caregivers (family members or close friends), and 197 oncologists confirmed the high prevalence and the adverse impact of fatigue in the population with cancer [24].

The foregoing results are all derived from studies with adults. Little is known about fatigue as a symptom in children and adolescents with cancer. So far, only Hockenberry-Eaton and colleagues [25] have explicitly evaluated fatigue as a symptom in these patients using a focus-group approach. Findings showed that the description of fatigue varies depending on the developmental level of the participants. Adolescents used a larger vocabulary to describe what fatigue is like and discriminated between mental and physical fatigue. A reduction in activity and participation in sports and play are common in younger children who are experiencing fatigue. Mental descriptions such as feeling sad or mad may be a reflection of fatigue.

Several causes of fatigue described by children and adolescents in this study are found in the adult literature. The causes included treatment with chemotherapy, radiation therapy and surgery, as well as side-effects such as low blood counts and fever. Moreover, the adolescent groups described symptoms of worry and concern that resulted in fatigue. Children and adolescents identified interventions that helped to decrease fatigue such as rest or naps, having fun, going outside, protected rest time and keeping busy.

To the best of our knowledge, no other studies explicitly focusing on fatigue are found in the literature on childhood cancer, but this does not mean that fatigue does not exist. Bottomley and colleagues [26] evaluated an instrument developed to measure childhood cancer stressors. They found that over 50% of a group of 75 school-age children with cancer reported being tired, not sleeping well and being unable to do the things they wanted to do. More than half of the children were not as active as before the illness, and reported to play less.

Information about off-treatment fatigue is rare in both adult and childhood cancer patients. Most of what is known about off-treatment fatigue stems from studies which have assessed fatigue in the context of a wider investigation on psychological adjustment or quality of life after cancer treatment. These studies suggest that significant degrees of fatigue may persist for years after the treatment ends, even if all known causes have been resolved and the patients are in a stage of remission [27-30]. For example, Fobair and colleagues [27] surveyed 403 patients who received radiation treatment alone (38%), radiation treatment plus chemotherapy (58%), or chemotherapy alone (5%) for treatment of adult and paediatric Hodgkin's disease. The median time since treatment was 9 years, 90% of the persons reported an adverse effect of treatment on the energy level and 37% felt that the energy level had not returned to normal.

A few studies, which focus on quality of life and psychosocial adjustment in long-term survivors of childhood cancer show that fatigue is a symptom experienced in this group of people. Wasserman and colleagues [31] examined the psychological status of 40 survivors of childhood and adolescent Hodgkin's disease. Treatment had ended 7 to 19 years prior to the study. One of the physical residual effects, reported by 5% of the survivors, was easy fatigability. Kanabar and colleagues [32] assessed late effects in a group of 30 survivors of childhood cancer who had completed treatment with chemotherapy and/or radiation therapy followed by autologous bone marrow rescue. The mean time lapsed since completion of therapy was 4 years. Four out of 50 patients expressed problems related to lack of energy. Finally, Moe and colleagues [33] examined the long-term survival and the quality of life of 93 adults diagnosed with acute lymphoblastic leukaemia (ALL) in childhood and compared this with corresponding controls in the family. Patients were off all cancer treatment for at least 15 years. The authors found that the somatization score on the General Health Questionnaire with items closely related to fatigue demonstrated a significantly higher score for the ALL survivors than for the controls.

Because of the current lack of knowledge about fatigue in long-term childhood cancer survivors, more information on the prevalence and the phenomenology of off-treatment fatigue in this group is desirable. A better understanding of off-treatment fatigue in cancer survivors is essential to help patients with fatigue and to develop intervention strategies to ameliorate these symptoms. Therefore, we decided to explore the concept of fatigue from the perspective of the long-term survivor of childhood cancer. We hope that this will be a step forward in understanding the actual experience of fatigue and in generating theoretical knowledge.

Methods

A qualitative approach was used. Such methods are most suitable when the focus of a study is on particular experiences where little is known [34].

Sample
Since approximately 1968, the Emma Kinderziekenhuis/Academic Medical Center in Amsterdam has offered facilities for the systematic care of children with cancer. Indefinite surveillance of survivors has been a long-standing policy. Since 1996, a model for providing comprehensive care to the survivors has been developed and an out-patient clinic was established for all former childhood cancer patients treated in the Emma Kinderziekenhuis/Academic Medical Center, who are at least 18 years old and who are off treatment for more than 5 years. About 98% of all eligible survivors are screened every year according to standard protocols by a paediatric oncologist, an internist-oncologist and a paediatric oncology research nurse for late medical and psychosocial effects. At the time of the visit, the paediatric oncologist meets the survivor and introduces the concepts of follow-up care, systematic evaluation and late effects of treatment. The patients' history since the last follow-up visit is taken and a physical examination is performed by the internist-oncologist. Following this visit, there is a meeting with the research nurse in which the psychosocial issues are discussed.
During medical screening the paediatric oncologist fills in a checklist which was developed to measure complaints and disease/treatment related late effects currently experienced by the survivor. One of the items on the checklist is fatigue. Survivors were regarded as suffering from extreme fatigue if they indicated that extreme fatigue was present and interfered with their daily life. After screening the first 455 survivors, 18% of them (n=83) seemed to be extremely fatigued. During psychosocial screening between August 1998 and January 1999 a purposeful sample of 35 fatigued survivors was recruited personally by the research nurse who asked the survivors to take part in this study. No one refused to participate.

Procedure

After obtaining verbal consent a semi-structured interview was conducted. The research nurse used a list of questions derived from the study by Smets and colleagues [35]. The following data were collected during the interview: description of fatigue and symptoms, frequency and course of fatigue, impact of fatigue on daily life, factors worsening or relieving fatigue, hours and pattern of sleep and onset of fatigue. All survivors, interviewed individually by one and the same researcher, had to answer the same open-ended questions. The goal of the interview was to obtain the most detailed description possible about the experience and to ensure that the participants themselves, and not the interviewer, determined the discussed content. The investigator used probes such as "please tell me more about that" and "what does that mean to you?" to encourage clarification of the questions and in-depth descriptions. Each interview lasted between 30 and 45 minutes. During the interview data were collected by the research nurse through recording field notes of the responses and feelings of the survivors.

Data analysis

Two researchers independently analysed these data. They screened data several times to identify relevant sentences, phrases, or anecdotes. After extracting these phrases they coded the field notes and classified the data into properties, categories and themes. After analysing the data of the same first 10 survivors, the researchers verified their interim results by discussing the categories. This procedure was repeated twice until all data were analysed. Most of the time the researchers agreed on the analysis of the descriptions. The few instances of nonagreement were differences in nuances and were resolved by a short discussion. To further ensure trustworthiness and credibility one of the investigators presented the findings to five other fatigued childhood cancer survivors and asked them whether the findings represented the essence of their experiences. No changes or alterations were made after the original analysis.

As a consequence of the qualitative approach no percentages will be reported in the results section. The following terms will be used to give an indication of the prevalence of the different replies; "some" or " a few" for less than half of the survivors, and "most", or "the majority" for more than half of the survivors.

Results

Sample

The mean age of the persons was 27 years (range 18-38 years), 71% were women and 29% men. Fourty two percent of the persons were single, 55 % were married or lived with a partner, one person was divorced.

The mean age at the time of diagnosis was 7 years (range 1-19 years), the average time since the end of treatment was 17 years (range 8-25 years). The diagnoses included acute lymphoblastic leukaemia and lymphoma (n=15), nephroblastoma (n=6), brain tumors (n=4), and other solid tumors (n=10). Seventy four percent of the whole group had received chemotherapy, 66% radiotherapy, 60% underwent major surgery and 20 % had all three treatments. The survivors reported at least one ongoing physical problem related to the treatment. Most common problems were alopecia, infertility, soft tissue hypoplasia, endocrine dysfunction, and neuropsychological problems. Besides the possible influence of these conditions, no other causes for fatigue were found.

Description of fatigue and symptoms

Although most respondents found it very difficult to give a clear description, a variety of phrases came up. Both physical and mental components of fatigue were described, such as *"tired, just tired"* or *"apathetic, absent-minded"*. Others stated that they were *"exhausted"* or *"completely wrung out"*. Additionnally, they often used metaphors for fatigue such as feeling like *"a sort of slumber in my head"* or *"having rag doll legs"*. They frequently described their fatigue in relation to sleep or level of energy. Some people reffered to *"an irresistible urge to sleep and cannot keep my eyes open"*. One person commented *"sometimes I think that I just need a good kick under my ass before I can do something. I can't get going, I have to drag myself forward"*. Although some respondents mentioned a lack of energy as well as a lack of drive, this was certainly not true for all of them. Some persons stated that *"I do want to, but I cannot"*. Many experienced fatigue as if they exerted themselves to the utmost. Their responses were: *"As if you go on for 36 hours, then fall into a deep sleep for 5 minutes, and next, when you wake up, it is if you're broke"* and *" It's an effort for me to get through the day. When I wake up in the morning, it is as if the night already falls."*

These young survivors experienced a wide variety of other symptoms associated with fatigue, such as trembling, muscular pain, night sweats, and fever. Sleepiness, headache and poor eyesight were by far the most reported symptoms.

Frequency and course of fatigue

More than half of the respondents were fatigued *"most of the time"*. The others found that fatigue was present *"all the time"*. The course of fatigue during the day differed among the survivors, although the majority reported to be fatigued when waking up in the morning. For some respondents the fatigue was worst at the end of the afternoon or in the evening. Three respondents reported that the moment of utmost fatigue differed from day to day. This was illustrated by the remark of a 24-year-old survivor, *"It can hit me in the morning and then I'm not able to do as much as I would like to do. Sometimes I'll have it at the end of the afternoon and I have to lie down for hours. Then cooking a dinner is too much for me"*.

Impact of fatigue on daily life

Fatigue was defined by all respondents as a negative impact on their daily lives. They gave many examples of how fatigue limited their social activities or how they had to confine their activities to the essentials. As one 20-year-old survivor reported: *"I am not able to do as much as I want to do. I have no energy for pleasant things and I must cancel many appointments because I'm too tired. My social life is a mess"*.

A few married survivors described the consequences of fatigue for their spouse and children, as one woman stated, *"Sometimes, because of my fatigue, I cannot give enough attention to my kid. I feel very guilty when I don't play with her. And when my husband comes home from work, he ought to appreciate that the dinner is ready. After dinner I need his help to get the dishes done and to put our daughter to bed".*

The impact of fatigue extended to the work setting or the school. Survivors often described having less energy to continue their work or study. Four survivors were partly officially declared unfit for work because of fatigue. Six persons chose to have a part-time job instead of working full-time. As one survivor said, *"It is one thing or the other. I want to save my energy a bit for other things".*

A number of the survivors indicated that fatigue negatively affected their mood, some found the fatigue very discouraging and now and then had depressive feelings.

Many young respondents said that they planned activities to decrease expenditure of energy. *"I don't have the energy to clean the bathroom in one time. So in the morning I'll do one side and in the afternoon I'll do the other side. In between I have to sit and relax. You probably understand that it takes a lot of effort to run my household".*

Respondents were asked to compare their experience of fatigue with those of peers. Most survivors found their friends less fatigued, more active and more capable to do the things they liked, such as going to the movies or having an evening out. In the words of one survivor: *"They are tired after a busy day, I have done nothing and I'm tired already".*

What does help with fatigue and what does not help
About one third of the respondents could not think of any activity that might help to reduce the fatigue. The remainder described a variety of activities that helped to decrease fatigue. Sleep, rest and recreational activities were amongst the most frequently reported activities. One respondent described how rest helps: *"I usually take a few naps every day and that makes me feel better".* Activities that kept a person busy and helped to distract from feeling tired were mentioned by some, such as embroidering or to work on leisure interests.

We explicitly asked whether the persons participated in regular physical activities. Although some said that they felt too tired to go in for sports, most of them reported doing some regular exercise such as aerobics or fitness. Most survivors indicated that it did not positively influence the fatigue but that they did benefit psychologically from regular exercise. One said: *"I feel more relaxed and healthy when I exercise".*

Sleep pattern
None of the survivors reported sleep problems, about half of them reported to sleep 7.5 to 9 hours per night, others slept more than 9 hours. Eight persons said that they needed 10 to 12 hours sleep per night. In general, the quality of sleep was good. Most survivors *"slept like a log"* and did not complain about problems with falling asleep or staying asleep.

Onset of fatigue
Respondents were asked to indicate the moment at which they first became aware of the fatigue. Two persons diagnosed with cancer in adolescence reported that

fatigue was already present as a symptom of their disease before the diagnosis. Nine survivors, mostly diagnosed with cancer in adolescence, identified fatigue as a significant side-effect of the treatment. As one survivor said: *"During radiation therapy I felt great fatigue and weakness. After my treatment was finished I felt better and had some more energy, but I still tire very easily. In fact I am tired most of the time"*.

Twelve persons mentioned that, as far as they could recall, they had suffered from fatigue their entire life. The majority of these survivors were toddlers or preschoolers at the time of cancer treatment.

The onset of fatigue was described differently by the remainder of survivors, some became aware of fatigue when they went to high school. One man observed fatigue at the time of his first job and had less energy than his colleagues and peers. One woman recognized fatigue as a symptom during pregnancy, this was 25 years after treatment for ALL. She reported that *"fatigue has been quite a problem since then"*. Now, 9 years after her delivery, she is still extreme fatigued. Finally, some survivors reported that although they never had been very active their energy level decreased.

Table 1 gives a summary of the key findings.

Table 1. Key findings of fatigue in childhood cancer survivors

---

- Fatigue is described as exerting oneself to the utmost, both physically and mentally
- Fatigue is present most or all of the time, since diagnosis and/or treatment
- Fatigue is frequently already present upon awakening, despite an average of 9 hours sleep per night
- Fatigue negatively affects daily and social activities

Discussion

To our knowledge our study is the first directed at the concept of fatigue in a sample of young adult survivors of childhood cancer. Findings revealed that fatigue is a serious problem for some persons and affects many aspects of their daily life. This is in line with the outcome of other studies in adult cancer patients suffering from fatigue during or following treatment [28,36-38]. It also appears that survivors adjust their activities to cope with the symptoms of fatigue. This finding was reported in another study about fatigue also in adult cancer patients [39].

Some aspects of our study need to be mentioned. A limitation was that the interviews were not audio-taped. Just making field notes can lead to investigator bias, which involves the tendency to selectively observe and record certain data at the expense of other data [40]. Such a bias can not be ruled out. In interpreting the field notes, however, an attempt was made to avoid further bias by having two researchers independently screen the notes. To test the validity of the findings, results were presented to five other extremely fatigued childhood cancer survivors.

In our study women are over-represented. Of all the survivors screened in our follow-up clinic so far, 18% reported to be extremely fatigued; 72% of this group of extremely fatigued survivors were women. This over-representation of women may reflect the fact that women in general report more fatigue than men [41,42]. The experience of fatigue in men and women seemed to be comparable. However, possible gender differences in the experience of fatigue have to be investigated systematically in a larger study.

Our findings rely on self-report. Fatigue, like pain, is a subjective experience and is thus never completely accessible without a certain degree of self-report. We asked the survivors to describe in retrospect their experience of fatigue, which might lead to recall bias. On one hand, it is possible that the survivors, by emphasizing fatigue during the interview, may tend to overreport their experience. On the other hand, the survivors might underreport their fatigue because they have already adjusted their daily activities to fatigue for such a long time. More than half of our respondents indicated that fatigue developed before their illness or during the treatment period, and some of the survivors thought that, as far as they knew, fatigue had been present during their whole life. These findings support the outcomes of Hockenberry-Eaton and colleagues [25], who suggest that fatigue is present in children with cancer. Some survivors became fatigued several years after the treatment. Changes in a daily pattern, e.g. going to high school, starting a job, or becoming pregnant, made the survivors aware of their fatigue. It is uncertain, however, whether fatigue was initiated by these changes. The question remains if the fatigue was really caused by their childhood cancer or by its treatment. It is of the utmost importance to investigate the prevalence of fatigue in a comparable young general population sample to ascertain whether survivors of childhood cancer differ, as far as fatigue is concerned, from healthy persons.

A striking finding is the amount of sleep time reported by the survivors. Although the preferred sleep duration for normal young adults varies, the average time is about 7.5 hours. Only 2% of the normal population reports sleeping more than 9 hours [43]. In our study group, we saw that 49% slept 9 hours or more. It is also striking that most of our childhood cancer survivors already felt fatigued on waking up, even after sleeping so many hours per night without sleeping impairments. Glaus [44] measured and explored fatigue in a group of adult cancer patients, non-cancer patients and healthy individuals. The results indicated that healthy individuals reported that they felt fit in the morning and tired after work, with steadily increasing levels of fatigue over the day. In the cancer patient group, the daily profile was different: fatigue was continuously present, patients already felt fatigued in the morning and, to a certain degree, over the whole day. These findings partly correspond with our results.

Another point is that the relation in our study group between fatigue and depression is unclear. One of the features of depression is being fatigued when waking up in the morning. Since the majority of our survivors reported to be fatigued already in the morning, one might expect them to be depressed as well. Further research is required to investigate the relation between fatigue and depression in childhood cancer survivors.

Although our study clearly suggests the existence of fatigue in a number of childhood cancer survivors, the psycho-biological processes which potentially underly off-treatment fatigue remain unclear. Further examination must also concentrate on the epidemiology of fatigue. It is essential to establish the prevalence of fatigue among childhood cancer survivors to know the scope of fatigue and to identify persons who are most at risk. Future research should also focus on the pathophysiological mechanisms of fatigue in survivors.

Finally, it is crucial that health-care providers acknowledge the impact of fatigue on the quality of life in cancer patients and that they give it a closer attention and evaluation and, whenever possible, therapeutic intervention [45]. In the study by Vogelzang and colleagues [24], almost one third of the patients mentioned fatigue to their oncologist at every visit and only 6% of oncologists believed that it was mentioned that often. Paradoxically, 80% of oncologists believed that fatigue is overlooked or undertreated. The authors concluded that the observations strongly suggest that patients and oncologists find it difficult to communicate about fatigue.

Although the findings of qualitative studies are not meant to be generalized, our findings should be kept in mind by nurses and other health-care providers in oncology follow-up clinics. As some childhood cancer survivors are now recognized to possibly be at risk of fatigue, we recommend the routine screening of all survivors for this symptom. As Piper [46] suggested, health-care providers can ask survivors three simple questions to assess the severity of the fatigue:

• Are you experiencing any fatigue?
• If yes, how severe is the fatigue?
• How is the fatigue interfering with the activities of daily living?

If a patient indicates symptoms of fatigue, nurses and other health-care providers can use this knowledge to plan interventions to decrease the fatigue or help the patient to live with fatigue. Possible interventions might be energy conservation activities, a planned exercise programme, stress reduction techniques or a nutritional counselling programme.

In conclusion, survivors were very glad with our commitment to investigating this symptom and welcomed the opportunity to talk. In the words of one survivor, *"I'm very glad that you acknowledge fatigue because most of the time it's like, well, I felt a bit guilty, as if I was the cause and now, I feel as if it is okay".*

Acknowledgement
We are grateful to the people willing to participate in our study.

References

1. Blatt J, Copeland DR, Bleyer WA. Late effects of childhood cancer and its treatment. In: Pizzo PA , Poplack DG, eds. Principles and Practice of Pediatric Oncology. JB Lippincott Company 1993, 1303-1329.
2. Hobbie W, Ruccione K, Moore IK, Truesdell S. Late effects in long-term survivors. In: Foley GV, Fochtman D, Mooney KH, eds. Nursing Care of the Child with Cancer. Orlando, Florida: W.B. Saunders Company 1993, 466-496.
3. Schwartz CL, Hobbie WL, Constine LS. Survivors of Childhood Cancer: Assessment and management. St. Louis, Missouri: Mosby-Year Book Inc 1994.
4. Fletcher JM, Copeland DR. Neurobehavioral effects of central nervous system prophylactic treatment of cancer in children. *J Clin Exp Neuropsychol* 1988, 10, 495-538.
5. Tucker MA, Meadows AT, Boice JD, Hoover RN, Fraumeni JF. Cancer risk following treatment of childhood cancer. In: Boice JD, Fraumeni JF, eds. Radiation Carcinogenesis: Epidemiology and Biological Significance. New York: Raven, 1984, 211-224.
6. Glaus A, Crow R, Hammond S. A qualitative study to explore the concept of fatigue/tiredness in cancer patients and in healthy individuals. *Eur J Cancer Care* (English Language Edition) 1996, 5, 8-23.
7. Piper BF. Fatigue. In: Carrieri, Lindsey , West, edis. Pathophysiological Phenomena in Nursing. Human Responses to Illness. Philadelphia: W B Saunders Company 1986.
8. Aistars J. Fatigue in the cancer patient: a conceptual approach to a clinical problem. *Oncol Nurs Forum* 1987, 14, 25-30.
9. Cimprich B. Attentional fatigue following breast cancer surgery. *Res Nurs Health* 1992, 15, 199-207.
10. Grant M, Ferrell BR, Schmidt G, Fonbuena P, Niland JC, Forman SJ. Measurment of quality of life in bone marrow transplantation survivors. *Qual Life Res* 1992, 1, 375-384.
11. Blesch KS, Paice JA, Wickham R et al. Correlates of fatigue in people with breast or lung cancer. *Oncol Nurs Forum* 1991, 18, 81-87.
12. Nail LM, Jones LS, Greene D, Schipper DL, Jensen R. Use and perceived efficacy of self-care activities in patients receiving chemotherapy. *Oncol Nurs Forum* 1991, 18, 883-887.
13. Pickard-Holley S. Fatigue in cancer patients. A descriptive study. *Cancer Nurs* 1991, 14, 13-19.
14. Graydon JE, Bubela N, Irvine D, Vincent L. Fatigue-reducing strategies used by patients receiving treatment for cancer. *Cancer Nurs* 1995, 18, 23-28.
15. Haylock PJ, Hart LK. Fatigue in patients receiving localized radiation. *Cancer Nurs* 1979, 2, 461-467.
16. Greenberg DB, Sawicka J, Eisenthal S, Ross D. Fatigue syndrome due to localized radiation. *J Pain Symptom Manage* 1992, 7, 38-45.
17. Irvine D, Vincent L, Graydon JE, Bubela N, Thompson L. The prevalence and correlates of fatigue in patients receiving treatment with chemotherapy and radiotherapy. A comparison with the fatigue experienced by healthy individuals. *Cancer Nurs* 1994, 17, 367-378.
18. Piper BF, Rieger PT, Brophy L, Haeuber D, Hood LE, Lyver A, Sharp E. Recent advances in the management of biotherapy-related side effects: fatigue. *Oncol Nurs Forum* 1989, 16, 27-34.
19. Dean GE, Spears L, Ferrell BR, Quan WD, Groshon S, Mitchell MS. Fatigue in patients with cancer receiving interferon alpha. *Cancer Pract* 1995, 3, 164-172.
20. Richardson A, Ream E. The experience of fatigue and other symptoms in patients receiving chemotherapy. *Eur J Cancer Care* (English Language Edition) 1996, 5, 24-30.
21. Stone P, Richards M, Hardy J. Fatigue in patients with cancer. *Eur J Cancer* 1998, 34, 1670-1676.
22. Mermelstein HT, Lesko L. Depression in patients with cancer. *Psychooncology* 1992, 1, 199-215.
23. Visser MR, Smets EM. Fatigue, depression and quality of life in cancer patients: how are they related? *Support Care Cancer* 1998, 6, 101-108.
24. Vogelzang NJ, Breitbart W, Cella D, Curt GA, Groopman JE, Horning SJ, Itri LM, Johnson DH, Scherr SL, Portenoy RK. Patient, caregiver, and oncologist perceptions of cancer-related fatigue: results of a tripart assessment survey. The Fatigue Coalition. *Semin Hematol* 1997, 34, 4-12.
25. Hockenberry-Eaton M, Hinds PS, Alcoser P, O'Neill JB, Euell K, Howard V, Gattuso J, Taylor J. Fatigue in children and adolescents with cancer. *J Pediatr Oncol Nurs* 1998, 15, 172-182.

26. Bottomley S, Teegarden C, Hockenberry-Eaton M. Fatigue in children with cancer. Clinical considerations for nurses. *J Pediatr Oncol Nurs* 1995, 13, 178.
27. Fobair P, Hoppe RT, Bloom J, Cox R, Varghese A, Spiegel D. Psychosocial problems among survivors of Hodgkin's disease. *J Clin Oncol* 1986, 4, 805-814.
28. Belec RH. Quality of life: perceptions of long-term survivors of bone marrow transplantation. *Oncol Nurs Forum* 1992, 19, 31-37.
29. Ferrell BR, Dow KH, Leigh S, Ly J, Gulasekaram P. Quality of life in long-term cancer survivors. *Oncol Nurs Forum* 1995, 22, 915-922.
30. Mast ME. Correlates of fatigue in survivors of breast cancer. *Cancer Nurs* 1998, 21, 136-142.
31. Wasserman AL, Thompson EI, Wilimas JA, Fairclough DL. The psychological status of survivors of childhood/adolescent Hodgkin's disease. *AJDC* 1987, 141, 626-631.
32. Kanabar DJ, Attard-Montalto S, Saha V, Kingston JE, Malpas JE, Eden OB. Quality of life in survivors of childhood cancer after megatherapy with autologous bone marrow rescue. *Ped Hemat Oncol* 1995, 12, 29-36.
33. Moe PJ, Holen A, Glomstein A, Madsen B, Hellebostad M, Stokland T, Wefring KW, Steen-Johnson J, Nielsen B, Howlid H, Borsting S, Hapnes C. Long-term survival and quality of life in patients treated with a national all protocol 15-20 years earlier: IDM/HDM and late effects? *Ped Hemat Oncol* 1997, 14, 513-524.
34. Morse JM. Qualitative Nursing Research. A Contemporary Dialogue. Newbury Park, California: Sage Publications 1991.
35. Smets EMA, Visser MRM, Willems-Groot AFMN, Garssen B, Oldenburger F, van Tienhoven G, de Haes JCJM. Fatigue and radiotherapy: (A) experience in patients undergoing treatment. *Br J Cancer* 1998, 78, 899-906.
36. Ferrell BR, Grant M, Dean GE, Funk B, Ly J. "Bone tired": The experience of fatigue and its impact on quality of life. *Oncol Nurs Forum* 1996, 23, 1539-1547.
37. Smets EMA, Visser MRM, Willems-Groot AFMN, Garssen B, Schuster-Uitterhoeve ALJ, de Haes JCJM. Fatigue and radiotherapy: (B) experience in patients 9 months following treatment. *Br J Cancer* 1998, 78, 907-912.
38. Irvine DM, Vincent L, Graydon JE, Bubela N. Fatigue in women with breast cancer receiving radiation therapy. *Cancer Nurs* 1998, 21, 127-135.
39. Rhodes VA, Watson PM, Hanson BM. Patients' descriptions of the influence of tiredness and weakness on self-care abilities. *Cancer Nurs* 1988, 11, 186-194.
40. Hunter DE, Foley MB. Doing anthropology: A student centered approach to cultural anthropology. New York: Harper & Row 1976.
41. Lewis G, Wessely S. The epidemiology of fatigue: more questions than answers. *J of Epidemiol Comm Health* 1992, 46, 92-97.
42. Smets EMA. Fatigue in cancer patients undergoing radiotherapy. 1997. Thesis. Amsterdam, The Netherlands.
43. Baker TL. Introduction to sleep and sleep disorders. *Med Clin North Am* 1985, 69, 1123-1152.
44. Glaus A. Assessment of fatigue in cancer and non-cancer patients and in healthy individuals. *Support Care Cancer* 1993, 1, 305-315.
45. Groopman JE. Fatigue in cancer and HIV/AIDS. *Oncology* 1935, 12, 335-344.
46. Piper BF. The Groopman article reviewed. *Oncology* 1998, 12, 345-346.

Chapter 5
No excess fatigue in young adult survivors of childhood cancer

Abstract

Purpose: Clinical reports suggest that many survivors of childhood cancer experience fatigue as a long-term effect of their treatment. To investigate this issue further, we assessed the level of fatigue in young adult survivors of childhood cancer. We compared the results with a group of young adults with no history of cancer. The impact of demographic, medical and treatment factors and depressive symptoms on survivors' fatigue was studied.

Patients and methods: Participants were 416 long-term survivors (LTS) of childhood cancer (age range 16-49 years, 48% of whom were female) who had completed treatment an average of 15 years previously and 1026 persons (age range 16-53 years, 55% female) with no history of cancer. All participants completed the Multidimensional Fatigue Inventory (MFI-20), a self-report instrument consisting of five scales (general fatigue, physical fatigue, mental fatigue, reduced activity, reduced motivation) and the Center for Epidemiologic Studies Depression Scale (CES-D).

Results: Small differences were found in mean scores for the different dimensions of fatigue between the LTS group and controls (range effect sizes -0.34 to 0.34). Women experienced more fatigue than men. Multiple linear regression analysis revealed that being female and unemployed were the only demographic characteristics explaining the various dimensions of fatigue. With regard to medical and treatment factors, diagnosis and severe late effects/health problems were associated with fatigue. Finally, depression was significantly associated with fatigue on all subscales.

Conclusion: Our clinical practice suggests a difference in fatigue in young adult childhood cancer survivors and their peers. This could not be confirmed in this study using the MFI-20. The well known correlation between depression and fatigue was confirmed in our study. Further research is needed to clarify the undoubtedly complex somatic and psychological mechanisms responsible for the development, maintenance and treatment of fatigue in childhood cancer survivors.

Introduction

Since the 1970s, the results from the treatment of children with cancer have significantly improved. Nowadays, the cure rate is about 70% and in countries with advanced medical care 1 in every 1000 young adults are cured [1]. It is well recognised that treatment can be associated with significant adverse late effects, including second malignancies, endocrine abnormalities, cardiac dysfunction, pulmonary disease, hepatic, renal, and gastrointestinal dysfunction, neurocognitive dysfunction, and psychologic sequelae [2-4]. While these long-term effects are discussed extensively in the literature, there is limited discussion of fatigue as a symptom experienced by survivors of childhood cancer.

Among cancer patients fatigue is a common, debilitating, and distressing symptom, due to their illness and/or their treatment [5-8]. While different definitions of cancer-related fatigue have been suggested, most include references to tiredness [9] or weariness, weakness, exhaustion and lack of energy [8,10]. Recent multidimensional conceptualisations of fatigue in adult cancer patients suggest that fatigue is a subjective experience with significant physical (e.g. weakness), behavioural (e.g. alterations in sleep patterns and activity level), cognitive, and affective (e.g. mood disturbance) components [7-9,11]. Fatigue can have serious adverse effects on quality of life [12,13], as well as a considerable impact on self-care activities [14].

It has also been reported that some patients experience significant degrees of fatigue long after the conclusion of cancer treatment, even if all known causes have been resolved and the patients are in remission [15-31]. These studies investigated patients treated for breast cancer, Hodgkin's disease, lymphoma patients, and various types of cancer. The mean time elapsed since treatment varied from 9 months to 12 years and percentages of fatigue from 17% to 30% [32].

To our knowledge, no studies in the literature have specifically focused on fatigue in childhood cancer survivors. Wasserman and colleagues [33] reported "easy fatigability" in 5% of 40 survivors of childhood and adolescent Hodgkin's disease. In the study by Kanabar and colleagues [34], four out of 50 survivors who had completed treatment with chemotherapy and/or radiation therapy followed by autologous bone marrow rescue had problems related to a lack of energy. Moe and colleagues [35] found that the somatization score on the General Health Questionnaire with items closely related to fatigue demonstrated a significantly higher score for the survivors of acute lymphoblastic leukaemia (ALL) than for the controls. Finally, Zeltzer and colleages [36] reported no difference between the Profile of Moods State Fatigue subscale score of 552 survivors of childhood ALL and 394 sibling controls. Because no difference in fatigue was found in this population, a further evaluation of risk factors for fatigue was not performed. The above findings contrast to clinical observation and reports of former childhood cancer patients. Results from a former qualitative study performed by us, indicate that fatigue is a serious problem in a subgroup of young adult survivors of childhood cancer and that fatigue affects many aspects of quality of life [37].

However, the symptom of fatigue is not specific for cancer. Fatigue and lack of energy is a prevalent symptom in the general population, and prevalence estimates range from 11% to 45% [38,39]. Fatigue is also a major complaint among attenders of general practices [40,41], and it is a central symptom in many other diseases, for example, ischaemic heart disease [42] and depression [41]. To interpret the significance of results obtained in follow-up studies involving childhood cancer patients, a comparison should therefore be made with persons without a history of cancer. Some studies indicate that disease-free cancer patients report more fatigue than controls do [15,18,19,21,22]. However, in other studies no differences in fatigue scores were found between cancer patients, following treatment 22 and 9 months before, respectively, and healthy comparisons [17,20].

Little is known about the role of demographic, medical and former treatment modalities in predicting post-treatment fatigue. Some studies have found higher prevalences or higher mean scores in women [17,38,39,41,43,44], whereas others did not find such differences [45,46]. Servaes and colleagues [32] evaluated 16 studies in which the focus was on "off-treatment fatigue". Most studies found no relationships between present fatigue and former disease- and treatment variables. In two studies, the severity of post-treatment fatigue was related to the extent of treatment. In these studies, former chemotherapy patients (sometimes in combination with radiation therapy and/or hormonal therapy) reported higher levels of fatigue compared with those treated with radiation therapy alone [16,17,19,20,24].

The relationship between depression and fatigue experienced by cancer survivors has not been systematically studied *per se*; however, most of the data suggests a positive relationship between depressive symptoms and fatigue [15,17-20,31]. The relationship between these two constructs is clearly complex. Fatigue may be the result of a depressed mood [47]. However, the person who continuously perceives his or her energy as insufficient may become depressed. To complicate matters, in cancer depression and fatigue may co-occur without having a causal relationship, because they can both originate from the same pathology [48].

The aim of the present study was to assess the level of fatigue in young adult survivors of childhood cancer. The severity was compared with that observed in a group of young adults with no history of cancer. Furthermore, the impact of demographic, medical and treatment factors and depressive symptoms on survivors' fatigue was studied.

Patients and methods

Study group
Data were collected from two samples: young adult survivors of childhood cancer (hereafter referred to as the survivors group) and a reference group of persons with no history of cancer (hereafter referred to as the comparison group).

Survivors group
Patients eligible for participation in the study were those attending the long-term follow-up clinic at The Emma Kinderziekenhuis/Academic Medical Center, Amsterdam for their annual evaluation between January 1997 and July 1999. The long-term follow-up clinic was established in 1996 to monitor long-term seque-lae of childhood cancer and its treatment. Patients become eligible for transfer from active-treatment clinics to the follow-up clinic when they had succesfully completed their cancer treatment at least 5 years earlier. Survivors are evaluated annually in the clinic by a paediatric oncologist (persons aged <18 years) or internist-oncologist (persons aged >18 years) for late medical effects, as well as a research nurse or psychologist for psychosocial effects. Study participants had to be aged 16 years or older, have had a pathologic confirmation of malignancy and their cancer had to have been diagnosed before the patients were 19 years of age. Within the study period a total of 459 patients were offered appointments to attend the follow-up clinic. Of these patients, 11 attended but were not included in this study (1 person was schizophrenic, 8 were developmentally delayed and 2 were deemed ineligible because of a current health problem causing emotional upset). A further 32 patients did not attend, representing a failed attendance rate

over this period of 7%. Survivors who did attend were significantly younger than survivors who did not attend (mean = 24 versus 26 years; P < 0.05), but there were no differences with regard to sex or type of diagnosis. A total of 416 patients were approached to take part in this study during their visit at the follow-up clinic and all agreed to participate. After their informed consent the survivors were individually asked to complete a questionnaire by one of the authors (NL). The investigator was present to make sure that the questionnaire was clearly understood. Most survivors had no difficulty in completing the questionnaire, only 2 persons needed some assistance.

Comparison group
Participants were recruited with the help of survivors' general practitioners (GPs). Letters with response cards were sent to the GPs of the potential survivor group (n=540) explaining the purpose of the study and asking for their help in selecting an age-matched control group. Three hundred and thirty GPs responded (61% response rate), of which 151 stated that they could not participate because of a lack of time: thus, 179 of the notified GPs agreed to take part in the study. These GPs were asked to select 10 subjects from their patients registry lists (starting with a given letter from the alphabet) with a given sex and age range. Those with (a prior history of) cancer were to be excluded. The GPs had to send to those who were eligible a packet containing the questionnaire, a stamped return address envelope and a cover letter. Two weeks after the original mailing date the GPs had to send another packet with the same content and a reminder letter. Of 1790 questionnaires mailed, 23 were returned because the subjects had moved and their address could not be traced. Twenty-four persons refused to participate for various reasons (lack of time n= 5; not able to understand Dutch n=4; perceived invasion of privacy n=7; other reasons n=8). Of the remaining 1743 questionnaires 1096 completed questionnaires were returned (response rate 63%). Four responses were excluded afterwards because the respondents did not met the inclusion criteria (too young n=3; too old n=1) and an additional 66 questionnaires lacked responses in many items. Thus, complete data were obtained from 1026 of those eligible. No significant differences were found with respect to age, sex, educational status and marital status in a Chi-square analysis to test for differences between controls who provided complete data and the 66 controls who did not.

Measures
Data were collected on sociodemographic characteristics in terms of age at follow-up, gender, marital status (single, living together/married), educational level (low = less than high school, high = high-school or advanced degree), and employment status (unemployed, student/homemaker, employed). In addition, a paediatrician, who had experience regarding the long-term effects of childhood cancer, reviewed the medical record prior to the visit to obtain information about survivor's cancer history and these data were recorded onto structured data coding sheets. Age at diagnosis, time since completion of therapy and duration of treatment were assessed. Diagnoses were categorised into: leukaemia/non-Hodgkin's lymphoma with or without cranial radiation therapy (CRT), solid

tumours, and brain/central nervous system (CNS) tumours. Treatment was aggregated into three categories: chemotherapy (with or without surgery), radiation therapy (with or without surgery), and combination therapy (chemotherapy and radiation therapy with or without surgery). Finally, late effects and health problems were scored on an adapted version of the Greenberg, Meadows & Kazak's Scale for Medical Limitations [49] by two paediatric oncologists and a paediatric oncology nurse, who were blinded to the survivors' identity. Patients were categorised into the following three groups according to their most serious medical limitation: 1 (mild) = no limitations of activity. This group included children with one kidney and second benign neoplasm's, and no cosmetic or organ dysfunction; 2 (moderate) = no serious restriction of daily life. This group included children with hypoplasia or asymmetry of soft tissue, mild scoliosis and other mild orthopaedic problems, moderate obesity, abnormally short stature, mild hearing loss, cataract, hypothyroidism, delayed sexual maturation, learning delay, enucleation of one eye, small testis, elevated follicle stimulating hormone/luteinising hormone (FSH/LH), alopecia, hypertension, pulmonary diffusion disturbances; 3 (severe) = significant restriction on daily activity or severe cosmetic changes. This group included children with learning delay requiring special education, soft tissue or bone changes that alter appearance, severe asymmetry, absent limb, dental reconstruction, gonadal failure, azoospermia, known sterility, blindness, organ damage that limits activity, second malignant neoplasm's, hemipareses, fatigue that affect daily and social activities. Sixteen percent (n =68) of survivors were categorised as mild, 46% (n = 189) as moderate, and 38% (n =159) as severe.

Fatigue was measured with the Multidimensional Fatigue Inventory (MFI-20), which is a self-report instrument consisting of 20 statements that can be rated on a five-point scale ranging from 'yes, that is true' to 'no, that is not true'. Items are combined to form five scales, including general fatigue, physical fatigue, mental fatigue, reduced activity and reduced motivation [50,51]. General fatigue refers to fatigue expressed by people in terms of statements like "I feel tired" and "I feel rested". Physical fatigue refers to physical sensations related to the feelings of tiredness. Mental fatigue refers to deficits in cognitive functioning, such as having difficulties concentrating. Finally, reduced activity covers not doing any useful activities and reduced motivation covers lack of motivation to initiate such activities. Higher scores indicate higher levels of fatigue. The MFI-20 has well-established levels of reliability and validity among cancer patients [17]. The reliability (i.e. internal consistency) in our study, as measured by Cronbach's alpha coefficient, ranged from 0.74 for the reduced motivation scale to 0.90 for the general fatigue scale.

Depression was measured with a part of the Center for Epidemiologic Studies Depression scale (CES-D) [52], which is a 20-item self-reporting scale, developed to measure depressive symptomatology in the general population. Respondents rate the degree to which they have experienced each depressive symptom during the past week on a four-point frequency scale (0 = rarely or none of the time; 3 = most or all of the time). To avoid any overlap in symptomatology with the MFI-20, we used this scale without the items of the domain 'somatic and retarded activity' (n=7) [48]. In this paper we refer to the questionnaire as the mood component of depression. Scores on this mood component could range from 0 to 39, with higher scores indicative of greater depressive symptomatology. In our study, Cronbach's alpha coefficient for this scale was 0.88.

Statistical analysis
The Statistical Package for Social Sciences (SPSS) Windows version 9.01 was used for all statistical analyses. Descriptive statistics were performed for all of the variables. Differences in demographic characteristics between survivors and the comparison group were analysed with Chi-square tests or Student's t-tests. Differences (corrected for sex and age) between the mean MFI-20 subscale scores of the survivors and the comparison group were tested with the Student's t-test. To examine the magnitude of these differences, effect sizes were calculated by dividing the difference between a given mean score of the survivor group and the mean score in the comparison group by the standard deviation of scores in the comparison group. An effect size of 0.20 is considered small, whereas effect sizes about 0.50 and 0.80 or greater are moderate and large, respectively [53].
Univariate relationships between survivors' MFI-20 scores and the mood component of the CES-D were assessed by Pearsons correlation coefficients. To investigate which variables predict survivors' fatigue, all variables were stepwise presented to a multiple linear regression model to assess their independent prognostic value. For three variables (employment status, diagnosis, and treatment) we created dummy variables and took the first category (unemployment, leukaemia/non-Hodgkin lymphoma without CRT, and chemotherapy with or without surgery) as reference for the analysis. First (Step 1), survivors' demographic characteristics were presented to the model. Second (Step 2), medical and treatment characteristics were entered, followed by the mood component of the CES-D (Step 3). Finally, the demographic, medical and treatment characteristics and the mood component of the CES-D were entered in the regression together (Step 4). With this strategy, the contribution of the separate steps show which variables in particular contribute to fatigue. Every model was repeated five times, for every fatigue subscale (general fatigue, physical fatigue, mental fatigue, reduced activity, and reduced motivation) separately. For each regression, the explained variance (R square) was determined.

Results

Characteristics of survivors and comparison group
Information about the demographic and medical characteristics of the survivor group and the comparison group is listed in Table 1. The survivors' age at diagnosis ranged from 0 to 19 years (median 8), the range of the time since completion of therapy was 5 to 33 years (median 15). The median duration of treatment was 16 months (range 0-170 months). The survivors were treated for a variety of cancers. The most frequent diagnoses were leukaemia, non-Hodgkin's lymphoma, Hodgkin's disease and Wilms' tumour. We compared the distribution of cancer diagnoses of our study population with the distribution of diagnoses of the survivors, who were known to be alive but were not yet seen in the follow-up

Table 1. Demographic and medical characteristics of the study group

Variable	Survivors (n= 416)		Comparison group (n= 1026)	P-value
Age at follow-up (years)				
Mean ± SD	24 ± 5.2		26 ± 5.1	<0.001*
Sex %				
Men	52		45	
Women	48		55	0.02#
Marital status %				
Single	72		46	
Living together/married	28		54	<0.001#
Educational status %				
Lower level	67		57	
Higher level	33		43	<0.001#
Age at diagnosis (years)				
Mean ± SD	8 ± 4.7			
Time since completion of therapy (years)				
Mean ± SD	15 ± 5.9			
Duration of treatment (months)				
Mean ± SD	16 ± 20.4			
Diagnosis	N	%		
Leukaemia/non-Hodgkin's lymphoma without CRT	116	28		
Leukaemia/non-Hodgkin's lymphoma with CRT	87	21		
Solid tumour	183	44		
Brain/CNS tumour	30	7		
Treatment				
Chemotherapy (with or without surgery)	197	47		
Radiation therapy (with or without surgery)	29	7		
Combination therapy (chemotherapy and radiation therapy with or without surgery)	190	46		
Medical limitations	N	%		
None/mild	68	16		
Moderate	189	46		
Severe	159	38		

SD, standard deviation; CRT, cranial radiation therapy; CNS, central nervous system
* t-test
# Chi-square

clinic. We found an over-representation of leukaemias and lymphomas in our study group. For logistic considerations many survivors with these diagnoses were seen in the follow-up clinic during the first years after the clinic was set up. The median age at follow-up of the survivors was 24 years (range 16-49 years), this was slightly younger than that of the comparison group (median 26; range 16-53 years). There were more male than female survivors, whereas the reverse was true for the comparison group. Furthermore, a Chi-square analysis showed that survivors and controls differed in terms of marital status and educational level. More survivors had never married compared with controls and more survivors had a lower educational level than the controls. However, because these could be typical features of the survivor group, univariate analysis of variance was performed to examine the influence of age at follow-up, sex, subject status, marital status, educational level, and the interaction of these variables on the different domains of fatigue. We found no significant effects of subject status and marital status and subject status and educational level on any of the dependent variables. Therefore, the influence of marital status and educational level on the dependent variables were not further accounted for.

Table 2. Mean (SD) scores for the MFI-20 for the survivors and the comparison group in relation to sex and age at follow-up (higher scores indicates more fatigue)

Scale	Males		Females		≤ 25 years		26-30 years		≥ 31 years	
	survivors (n=216)	controls (n=463)	survivors (n=200)	controls (n=563)	survivors (n=290)	controls (n=497)	survivors (n=80)	controls (n=458)	survivors (n=46)	controls (n=71)
General fatigue	7.5 (4.3)	8.8 (3.8)**	10.9 (5.2)	10.5 (4.5)	8.8 (4.9)*	9.6 (4.2)*	10.0 (5.3)	10.0 (4.4)	9.9 (5.6)	9.8 (4.8)
Physical fatigue	6.9 (3.8)	7.1 (3.2)	9.7 (4.7)*	8.8 (4.3)*	7.7 (4.1)	8.0 (3.9)	9.4 (4.8)*	8.1 (3.9)*	9.7 (5.4)	8.3 (4.0)
Mental fatigue	7.9 (4.5)	8.0 (3.7)	10.1 (4.9)**	8.7 (4.1)**	8.9 (4.8)	8.6 (3.9)	8.9 (4.6)	8.1 (3.9)	9.3 (5.2)	8.5 (3.9)
Reduced activity	6.9 (3.6)*	7.7 (3.2)*	8.6 (4.2)	8.0 (3.6)	7.4 (3.7)	7.8 (3.3)	8.6 (4.7)	8.0 (3.4)	8.3 (4.3)	7.9 (3.9)
Reduced motivation	6.1 (2.8)**	7.1 (2.9)**	7.3 (3.5)	7.3 (3.1)	6.4 (3.1)*	7.0 (2.9)*	7.1 (3.2)	7.4 (3.1)	7.6 (3.8)	7.9 (3.3)

* Statistically significant difference (p <0.05) in means by t-tests
** Statistically significant difference (p <0.001) in means by t-tests

Fatigue in survivors versus comparison group

Survivors scored significantly lower (i.e., reflecting less fatigue) on general fatigue (P <0.05, effect size -0.14) and reduced motivation (P <0.05, effect size -0.19) but statistically higher (i.e., reflecting worse fatigue) for mental fatigue (P <0.05, effect size 0.15) than controls. In Table 2 the mean scores and standard deviations for the different dimensions of fatigue are presented for the survivors and the comparison group in relation to sex and age at follow-up. In general, females experienced higher levels of fatigue on all subscales than men. Male survivors had significantly lower scores on general fatigue, reduced activity and reduced motivation than their peers (effect sizes -0.34, -0.25 and -0.34, respectively). Female survivors reported significantly more fatigue on the scales physical fatigue (effect size 0.21) and mental fatigue (effect size 0.34) in comparison with their female peers. Within the age groups, survivors 25 years old or younger scored significantly lower for general fatigue and reduced motivation (effect sizes -0.19 and -0.21, respectively). Survivors 26-30 years old reported more physical fatigue than controls (effect size 0.33).

Association between survivors' depression and fatigue

Prior to the simultaneous regression analysis, Pearson product moment correlation's for depression and fatigue were performed (see Table 3). Significant correlation's are shown between CES-D mood scores and the MFI-20 subscale scores.

Table 3. Pearson's correlation of the mood component of the CES-D and the MFI-20 scores for the survivors

CES-D mood	MFI-20
0.61**	General fatigue
0.58**	Physical fatigue
0.51**	Mental fatigue
0.57**	Reduced activity
0.61**	Reduced motivation

** Statistically significant differences (P <0.001)

Prediction of survivors' fatigue by demographic, medical and treatment characteristics and depression

Table 4 presents the results of predictors of fatigue identified by multivariate regression analysis at each step. With regard to the demographic variables (Step 1), female gender was the strongest prognostic factor of fatigue on all subscales. Physical fatigue and reduced motivation were explained by an older age at follow-up. General fatigue, physical fatigue and reduced activity were associated with unemployment.

With regard to the medical and treatment characteristics (Step 2), severe late effects/health problems were associated with fatigue on all of the subscales.

Table 4. Simultaneous linear regressions (Beta) for survivors' fatigue[a]

	GF	PF	MF	RA	RM
Step 1					
Demographic characteristics					
Sexe (female)	0.34**	0.32**	0.22**	0.21**	0.19**
Age at follow-up (years)	0.08	0.22**	0.04	0.11	0.16*
Marital status (married)	0.01	- 0.03	- 0.09	- 0.02	- 0.04
Educational level (higher level)	0.01	0.04	- 0.08	0.06	- 0.02
Employment status[c]					
Student/homemaker	- 0.24*	- 0.23*	- 0.13	- 0.18*	- 0.11
Employed	- 0.18*	- 0.13	- 0.15	- 0.06	- 0.23
Total R^{2b}	15%	16%	7%	7%	6%
Step 2					
Medical and treatment characteristics					
Age at diagnosis (years)	- 0.04	0.02	- 0.11	- 0.09	- 0.01
Diagnosis[d]					
Leukaemia/non-Hodgkin's lymphoma with CRT	- 0.12	- 0.19*	0.03	- 0.10	- 0.04
Solid tumour	- 0.03	- 0.07	0.03	0.01	- 0.05
Brain/CNS tumour	- 0.09	- 0.10	0.03	0.01	0.04
Duration of treatment (months)	0.05	0.05	- 0.01	0.04	0.02
Years since completion of therapy	- 0.02	0.06	- 0.08	- 0.01	0.05
Late effects/health problems	0.26**	0.25**	0.20**	0.23**	0.20**
Treatment[e]					
Radiation therapy (with or without surgery)	0.12	0.18*	0.12	0.04	0.07
Combination therapy (with or without surgery)	0.05	0.10	0.06	0.05	0.01
Total R^{2b}	9%	13%	7%	7%	7%
Step 3					
Mood component of the CES-D					
Depression	0.61**	0.58**	0.51**	0.58**	0.61**
Total R^{2b}	37%	34%	26%	33%	38%
Step 4					
Demographic, medical and treatment characteristics and depression					
Sexe (female)	0.19**	0.18**	0.08	0.04	0.00
Age at follow-up (years)	0.01	0.25	- 0.24	0.33	- 0.30
Marital status (married)	0.04	- 0.01	- 0.04	0.01	0.01
Educational level (higher level)	0.03	0.06	- 0.04	0.09*	0.04
Employment status[c]					
Student/homemaker	- 0.12	- 0.08	- 0.09	- 0.01	0.03
Employed	- 0.20*	- 0.18*	- 0.08	- 0.14*	- 0.06
Age at diagnosis (years)	0.06	- 0.08	0.21	- 0.25	0.38
Diagnosis[d]					
Leukaemia/non-Hodgkin's lymphoma with CRT	- 0.16*	- 0.22**	0.01	- 0.10	- 0.04
Solid tumour	0.02	- 0.06	0.05	0.04	- 0.01
Brain/CNS tumour	- 0.08	- 0.09	0.01	0.02	0.04
Duration of treatment (months)	0.02	0.02	- 0.03	- 0.02	0.01
Years since completion of therapy	0.02	- 0.15	0.23	- 0.29	0.45
Late effects/health problems	0.14*	0.14*	0.10*	0.11*	0.08
Treatment[e]					
Radiation therapy (with or without surgery)	0.01	0.08	0.06	- 0.04	0.01
Combination therapy (with or without surgery)	0.04	0.09	0.06	0.06	0.01
Depression	0.54**	0.51**	0.46**	0.56**	0.59**
Total R^{2b}	46%	45%	29%	39%	40%

Abbreviations: GF: general fatigue, PF: physical fatigue, MF: mental fatigue, RA: reduced activity, RM: reduced motivation, CRT: cranial radiation therapy, CNS: central nervous system

[a] Within each step variables are presented in order of selection, see also Methods section

[b] R is the percentage of the total variation of the dependent variable score that is explained by the independent variables together

[c] reference group = unemployment, [d] reference group = leukaemia/non-Hodgkin's lymphoma without CRT, [e] reference group = chemotherapy (with or without surgery)

* Statistically significant differences ($p < 0.05$) ** Statistically significant differences ($p < 0.001$)

Physical fatigue showed a negative association with diagnosis, meaning that survivors with leukaemia/non-Hodgkin's lymphoma without CRT suffered from more fatigue than survivors with leukaemia/non-Hodgkin's lymphoma with CRT. Physical fatigue was also explained by treatment (radiation therapy with or without surgery).

In Step 3, depression was associated with fatigue on all subscales. After entering both demographic, medical and treatment characteristics and the mood component of the CES-D into the model (Step 4), the results showed that depression was the strongest predictor of fatigue on all subscales. Both general fatigue and physical fatigue were further associated with females, unemployment, survivors who have had leukaemia/non-Hodgkin's lymphoma without CRT, and severe late effects/health problems. Mental fatigue was associated with severe late effects/health problems. Reduced activity was associated with higher educational level, unemployment and severe late effects/health problems. The selected characteristics explained only a moderate proportion of the variability (R^2) of the fatigue scores: 29%-46%.

Discussion

In this study, the level of fatigue among young adult survivors of childhood cancer was compared with a sample of young adults with no history of cancer. In addition, the relationship between the demographic, medical and treatment factors and depression on the survivors' fatigue were examined. The following discussion summarises the main findings, considers the clinical implications, and identifies several directions for future research.
The survivors and the comparison group differed on some aspects that possibly affected the comparisons. Significant differences were observed with regard to marital status and educational level. However, it is doubtful that these differences were influential because fatigue was not significantly related to marital status and educational level. The response rate among the controls was highly satisfactory for a mailed survey. Nonetheless, there are several limitations to the study. First, as with all survey studies, the likelihood of respondents being those particularly interested in the topic is high. Second, the respondents may be those who suffered from physical problems and were motivated to respond. Third, the findings of this study are limited by the heterogeneous patient groups with regard to the cancers involved, treatment regimens and the crude assessment categories used.
The results of the study suggest that young adult survivors of childhood cancer experience a level of fatigue that is more or less the same as that "normally" experienced by persons of about the same age. We were, to put it mildly, somewhat surprised by the results because the survivor group was expected to be more fatigued. The present study was inspired by clinical reports from a subgroup of childhood cancer survivors who attended the long-term follow-up clinic in our hospital and complained about extreme fatigue which had a negative impact on their daily lives. Findings from a qualitative study done by our team, revealed that fatigue was a serious problem for some persons and affected many aspects of their daily life [37].

As described earlier, previous investigations that included a non-cancer comparison group also found comparable fatigue ratings between cancer patients and controls [17,20], and high rates of complaints of chronic fatigue are found in the general population and primary care studies as well. Therefore, on the basis of our results, we can not conclude that fatigue following treatment for childhood cancer is actually no more severe than "normal" fatigue. Several factors have to be taken into account. First, the lack of differences between the survivors and controls may be due to the so-called 'response-shift'. Response shift refers to a theory that as a result of changes in a subject's health state, a person may undergo changes in internal standards, values or conceptualisations [54-56]. In our setting it could imply that the experience of fatigue for a relatively long period could have changed a fatigued survivors' standard of measurement concerning fatigue and as a result fatigue has been underreported. Previous research has documented that response shift may adversely affect the results of self-reported outcomes in clinical trials and other longitudinal studies [57]. Second, in this study, we did not assess either the type of fatigue or intensity of activity nor the characteristics of fatigue. In our qualitative study [37], survivors gave many examples of how fatigue limited their activities or how they had to confine their activities to the essentials. Therefore, it is possible that fatigued survivors limit their activities to such a degree that as a result, their fatigue does not exceed the level found in the comparison group. Further work is needed to substantiate these findings. Future research should also address survivors' characteristics associated with fatigue. The results of some studies provide evidence that patients with cancer experience fatigue that is different from the fatigue experienced by a "healthy" reference group [17,58-60]. For instance, a small exploratory study by Glaus et al. [59] compared patterns of fatigue in healthy workers and cancer patients. The healthy workers started the day without tiredness, remained fit until the late afternoon and were very fatigued in the evening. In the cancer patient group, the daily profile was different: fatigue was continuously present, patients already felt fatigued in the morning, and, to a certain degree, over the whole day. The authors proposed a circadian rhythm for fatigue in cancer patients. These findings partly correspond with the results in our qualitative study where most of our childhood cancer survivors reported that they already felt fatigued on waking up, even after sleeping many hours per night without sleeping impairments. Further work is needed between these and other correlates of fatigue, such as sleep disturbances and psychological distress. Third, we did not investigate the outcome of fatigue. If we want to demonstrate any difference in fatigue between cancer survivors and a comparison group the impact on quality of life, mobility, self-care, social isolation and role change needs to be explored in more detail.

When we examined the relationship between demographic, medical and treatment factors and depressive symptoms on survivors' fatigue we found that, among the demographic factors, female gender was significantly associated with severe fatigue on all of the subscales. This is in line with the outcomes of other studies [17,38,39,41,43,44]. As suggested by Akechi and colleagues [61], a possible explanation for the repeated finding of greater fatigue in women could be the gender difference in the perception of symptoms suggested by Gijsbers van Wijk & Kolk [62], who indicated that there are consistent sex differences in symptom reporting, with women having the higher rates.

Unemployment was associated with higher levels of general and physical fatigue and reduced activity. It should be noted, however, that the unemployed group was small (n=42) and more than half of these survivors were, partly or fully, officially declared unfit for work because of medical problems. It is not difficult to imagine that these persons have less energy and a decreased activity level.

We found some relation between medical factors and severe fatigue in this study. It is difficult to determine whether the present results are reliable because, to our knowledge, no prior study has addressed the associations between fatigue and these factors in childhood cancer survivors. It is unclear why survivors who have had leukaemia or non-Hodgkin's lymphoma without CRT were more likely to be generally and physically fatigued than survivors with other diagnoses. No satisfactory explanation is available for this finding.

It is not surprising that survivors with severe medical limitations had a higher risk of being fatigued than survivors with moderate and none/mild limitations, especially since survivors with severe medical limitations are more likely to have symptoms related to their disease. In one of the few multivariate studies of correlates of fatigue, Irvine and colleagues [58] found symptom burden to be an important independent predictor of fatigue levels. Another aspect that has to be taken into account is the fact that survivors suffering from fatigue that affected their daily and social activities were put into the severe medical limitation group. However, when we looked at the number of fatigued survivors in this category (n=66), we found that more than half of these survivors (n=36) were not put into this category because of their fatigue alone, but suffered from other severe medical limitations as well.

No relationship was found between other medical and treatment factors and severe fatigue, such as age at diagnosis, duration of treatment and years since completion of therapy. This lack of association is in line with the results described in the majority of studies [32].

In accordance with results in the literature, depression was substantially related to fatigue. As noted before, the depression-fatigue association is a very complex one. Fatigue may result from a depressive mood [47], but depression could also be a result of persistent feelings of fatigue and this may especially be the case when the treatment for cancer has ended some time ago. Although an association between depression and fatigue cannot be ruled out as an explanation for the fatigue experienced during and after treatment for cancer, it is clearly an incomplete description of the underlying process [32]. As stated by Smets and colleagues, it underlines that the role of depression should be taken into account when trying to alleviate fatigue [63] and that psychological support should form one aspect of a program for the management of fatigue. Trijsburg and colleagues [64] reviewed 22 studies that explored the effectiveness of psychological treatment for patients with cancer. This review concluded that "tailored counselling", where counselling and support were provided according to patients' needs, was effective not only in reducing distress and enhancing self-esteem but also for in reducing fatigue.

In conclusion, this report represents the first evaluation of fatigue in childhood cancer survivors. The lack of difference in fatigue between the survivors and the controls is noteworthy. However, in our opinion, one must be wary of concluding that survivors' fatigue is a trivial complaint. There is no doubt that many childhood cancer survivors face a multitude of overwhelming and de-energizing problems and "I just don't have any energy" is one of them.

Presently, we are sorely limited in what we can offer patients who are fatigued after treatment for cancer. In our opinion, it is critical that health care providers acknowledge fatigue in order to grant it legitimacy. From clinical practice, we found that simply asking about our survivors' fatigue, listening, and taking the problem seriously helps them to cope and adjust to the problem. To be able to offer more than that, we need to understand more about the nature and mechanisms of cancer-related fatigue. There is a need for studies that use prospective longitudinal designs to yield more definitive information about the incidence and aetiology of fatigue.

Acknowledgements
The authors would like to thank the survivors and the persons of the comparison group who generously contributed to this study.

References

1. Dreyer ZE, Blatt J, Bleyer A. Late effects of childhood cancer and its treatment. In Pizzo PA, Poplack DG, eds. Principles and Practice of Pediatric Oncology. Philadelphia: Lippincott Williams & Wilkins 2002, 1431-1461.
2. Hobbie W, Ruccione K, Moore IK, Truesdell S. Late effects in long-term survivors. In Foley GV, Fochtman D, Mooney KH, eds. Nursing Care of the Child with Cancer. Orlando, Florida: W.B. Saunders Company 1993, 466-496.
3. Schwartz CL. Late effects of treatment in long-term survivors of cancer. *Cancer Treat Rev* 1995, 21, 355-366.
4. Meister LA, Meadows AT. Late effects of childhood cancer therapy. *Curr Probl Pediatr* 1993, 23, 102-131.
5. Oberst MT, Hughes SH, Chang AS, McCubbin MA. Self-care burden, stress appraisal, and mood among persons receiving radiotherapy. *Cancer Nurs* 1991, 14, 71-78.
6. Irvine DM, Vincent L, Bubela N, Thompson L, Graydon J. A critical appraisal of the research literature investigating fatigue in the individual with cancer. *Cancer Nurs* 1991, 14, 188-199.
7. Smets EM, Garssen B, Schuster-Uitterhoeve AL, de Haes JC. Fatigue in cancer patients. *Br J Cancer* 1993, 68, 220-224.
8. Winningham ML, Nail LM, Burke MB, Brophy L, Cimprich B, Jones LS et al. Fatigue and the cancer experience: the state of the knowledge. *Oncol Nurs Forum* 1994, 21, 23-36.
9. Piper BF, Lindsey AM, Dodd MJ. Fatigue mechanisms in cancer patients: developing nursing theory. *Oncol Nurs Forum* 1987, 14, 17-23.
10. Aistars J. Fatigue in the cancer patient: a conceptual approach to a clinical problem. *Oncol Nurs Forum* 1987, 14, 25-30.
11. Piper BF, Rieger PT, Brophy L, Haeuber D, Hood LE, Lyver A et al. Recent advances in the management of biotherapy-related side effects: fatigue. *Oncol Nurs Forum* 1989, 16, 27-34.
12. Ferrell BR, Grant M, Dean GE, Funk B, Ly J. "Bone tired": The experience of fatigue and its impact on quality of life. *Oncol Nurs Forum* 1996, 23, 1539-1547.
13. Irvine DM, Vincent L, Graydon JE, Bubela N. Fatigue in women with breast cancer receiving radiation therapy. *Cancer Nurs* 1998, 21, 127-135.
14. Rhodes VA, Watson PM, Hanson BM. Patients' descriptions of the influence of tiredness and weakness on self-care abilities. *Cancer Nurs* 1988, 11, 186-194.
15. Hann DM, Jacobsen PB, Martin SC, Kronish LE, Azzarello, LM et al. Fatigue in women treated with bone marrow transplantation for breast cancer: a comparison with women with no history of cancer. *Support Care Cancer* 1997, 5, 44-52.
16. Woo B, Dibble SL, Piper BF, Keating SB, Weiss MC. Differences in fatigue by treatment methods in women with breast cancer. *Oncol Nurs Forum* 1998, 25, 915-920.
17. Smets EMA, Visser MRM, Willems G, AFMN, Garssen B, Schuster-Uitterhoeve ALJ et al. Fatigue and radiotherapy: (B) experience in patients 9 months following treatment. *Br J Cancer* 1998, 78, 907-912.
18. Andrykowski MA, Curran SL, Lightner R. Off-treatment fatigue in breast cancer survivors: a controlled comparison. *J Behav Med* 1998, 21, 1-18.
19. Broeckel JA, Jacobsen PB, Horton J, Balducci L, Lyman GH. Characteristics and correlates of fatigue after adjuvant chemotherapy for breast cancer. *J Clin Oncol* 1998, 16, 1689-1696.
20. Hann DM, Jacobsen P, Martin S, Azzarello L, Greenberg H. Fatigue and quality of life following radiotherapy for breast cancer: a comparative study. *J Clin Psychiatry in Med Sett* 1998, 1, 19-33.
21. Loge JH, Abrahamsen AF, Ekeberg O, Kaasa S. Hodgkin's disease survivors more fatigued than the general population. *J Clin Oncol* 1999, 17, 253-261.
22. Howell SJ, Radford JA, Smets EM, Shalet SM. Fatigue, sexual function and mood following treatment for haematological malignancy: the impact of mild Leydig cell dysfunction. *Br J Cancer* 2000, 82, 789-793.
23. Knobel H, Loge JH, Nordoy T, Kolstad AL , Espevik T, Kvaloy S et al. High level of fatigue in lymphoma patients treated with high dose therapy. *J Pain Symptom Manage* 2000, 19, 446-456.
24. Bower JE, Ganz PA, Desmond KA, Rowland JH, Meyerowitz, BE et al. Fatigue in breast cancer survivors: occurrence, correlates, and impact on quality of life. *J Clin Oncol* 2000, 18, 743-753.
25. Okuyama T, Akechi T, Kugaya A, Okamura H, Imoto S, Nakano T et al. Factors correlated with

fatigue in disease-free breast cancer patients: application of the Cancer Fatigue Scale. *Support Care Cancer* 2000, 8, 215-222.

26. Curt GA, Breitbart W, Cella D, Groopman JE, Horning SJ, Itri LM et al. Impact of cancer-related fatigue on the lives of patients: new findings from the Fatigue Coalition. *Oncologist* 2000, 5, 353-360.

27. Loge JH, Abrahamsen AF, Ekeberg, Kaasa S. Fatigue and psychiatric morbidity among Hodgkin's disease survivors. *J Pain Symptom Manage* 2000, 19, 91-99.

28. Servaes P, Van der Werf S, Prins J, Verhagen S, Bleijenberg G. Fatigue in disease-free cancer patients compared with fatigue in patients with Chronic Fatigue Syndrome. *Support Care Cancer* 2001, 9, 11-17.

29. Cella D, Davis K, Breitbart W, Curt G. Cancer-related fatigue: prevalence of proposed diagnostic criteria in a United States sample of cancer survivors. *J Clin Oncol* 2001, 19, 3385-3391.

30. Knobel H, Harvard-Loge J, Brit-Lund M, Forfang K, Nome O, Kaasa S. Late medical complications and fatigue in Hodgkin's disease survivors. *J Clin Oncol* 2001, 19, 3226-3233.

31. Fobair P, Hoppe RT, Bloom J, Cox R, Varghese A, Spiegel D. Psychosocial problems among survivors of Hodgkin's disease. *J Clin Oncol* 1986, 4, 805-814.

32. Servaes P, Verhagen C, Bleijenberg G. Fatigue in cancer patients during and after treatment: prevalence, correlates and interventions. *Eur J Cancer* 2002, 38, 27-43.

33. Wasserman AL, Thompson EI, Wilimas JA, Fairclough DL. The psychological status of survivors of childhood/adolescent Hodgkin's disease. *AJDC* 1987, 141, 626-631.

34. Kanabar DJ, Attard-Montalto S, Saha V, Kingston JE, Malpas JE, Eden OB. Quality of life in survivors of childhood cancer after megatherapy with autologous bone marrow rescue. *Ped Hemat Oncol* 1995, 12, 29-36.

35. Moe PJ, Holen A, Glomstein A, Madsen B , Hellebostad M, Stokland T et al. Long-term survival and quality of life in patients treated with a national all protocol 15-20 years earlier: IDM/HDM and late effects? *Ped Hemat Oncol* 1997, 14, 513-524.

36. Zeltzer LK, Chen E, Weiss R, Guo MD, Robison LL, Meadows AT et al. Comparison of psychologic outcome in adult survivors of childhood acute lymphoblastic leukemia versus sibling controls: a cooperative Children's Cancer Group and National Institutes of Health study. *J Clin Oncol* 1997, 15, 547-556.

37. Langeveld N, Ubbink M, Smets E. 'I don't have any energy': the experience of fatigue in young adult survivors of childhood cancer. *EJON* 2000, 4, 20-28.

38. Chen MK. The epidemiology of self-perceived fatigue among adults. *Preventive Med* 1986, 15, 74-81.

39. Lewis G, Wessely S. The epidemiology of fatigue: more questions than answers. *J of Epidem Comm Health* 1992, 46, 92-97.

40. Bates DW, Schmitt W, Buchwald D, Ware NC, Lee J, Thoyer et al. Prevalence of fatigue and chronic fatigue syndrome in a primary care practice. *Arch Intern Med* 1993, 153, 2759-2765.

41. Fuhrer R, Wessely S. The epidemiology of fatigue and depression: A French primary-care study. *Psychol Med* 1995, 25, 895-905.

42. Appels A, Mulder P. Fatigue and heart disease. The association between 'vital exhaustion' and past, present and future coronary heart disease. *J Psychosom Res* 1989, 33, 727-738.

43. Pater JL, Zee B, Palmer M, Johnston D, Osoba D. Fatigue in patients with cancer: results with National Cancer Institute of Canada Clinical Trials Group studies employing the EORTC QLQ-C30. *Support Care Cancer* 1997, 5, 410-413.

44. Kroenke K, Wood DR, Mangelsdorff D, Meier NJ, Powell JB. Chronic fatigue in primary care. Prevalence, patient characteristics, and outcome. *JAMA* 1988, 260, 929-934.

45. Cathebras PJ, Robbins JM, Kirmayer LJ, Hayton BC. Fatigue in primary care: prevalence, psychiatric comorbidity, illness behavior, and outcome. *J Gen Intern Med* 1992, 7, 276-286.

46. Wessely S, Chalder T, Hirsch S, Pawlikowska T, Wallace P, Wright DJM. Postinfectious fatigue: prospective cohort study in primary care. *Lancet* 1995, 345, 1333-1338.

47. Mermelstein HT, Lesko L. Depression in patients with cancer. *Psychooncology* 1992, 1, 199-215.

48. Visser MR, Smets EM. Fatigue, depression and quality of life in cancer patients: how are they related? *Supp Care Cancer* 1998, 6, 101-108.

49. Greenberg HS, Kazak AE, Meadows AT. Psychologic functioning in 8- to 16-year-old cancer survivors and their parents. *J Pediatr* 1989, 114, 488-493.

50. Smets EM, Garssen B, Bonke B, de Haes JC. The Multidimensional Fatigue Inventory (MFI) psychometric qualities of an instrument to assess fatigue. *J Psychosom Res* 1995, 39, 315-325.
51. Smets EM, Garssen B, Cull A, de Haes JC. Application of the multidimensional fatigue inventory (MFI-20) in cancer patients receiving radiotherapy. *Br J Cancer* 1996, 73, 241-245.
52. Radloff LS. The CES-D Scale: A self-report depression scale for research in the general population. *Appl Psychol Measurement* 1977, 1, 385-401.
53. Cohen J. Statistical power analysis for the behavioral sciences. New York: Academic Press 1977.
54. Breetvelt IS, Van Dam FSAM. Underreporting by cancer patients: the case of response-shift. *Soc Sci Med* 1991, 32, 981-987.
55. Sprangers MA. Response-shift bias: a challenge to the assessment of patients' quality of life in cancer clinical trials. *Cancer Treat Rev* 1996, 22, 55-62.
56. Schwartz CE, Feinberg RG, Jilinskaia E, Applegate JC. An evaluation of a psychosocial intervention for survivors of childhood cancer: paradoxical effects of response shift over time. *Psychooncology* 1999, 8, 344-354.
57. Sprangers MA, Van Dam FS, Broersen J, Lodder L, Wever L, Visser MR et al. Revealing response shift in longitudinal research on fatigue-the use of the thentest approach. *Acta Oncol* 1999, 38, 709-718.
58. Irvine D, Vincent L, Graydon JE, Bubela N, Thompson L. The prevalence and correlates of fatigue in patients receiving treatment with chemotherapy and radiotherapy. A comparison with the fatigue experienced by healthy individuals. *Cancer Nurs* 1994, 17, 367-378.
59. Glaus A. Assessment of fatigue in cancer and non-cancer patients and in healthy individuals. *Support Care Cancer* 1993, 1, 305-315.
60. Glaus A, Crow R, Hammond S. A qualitative study to explore the concept of fatigue/tiredness in cancer patients and in healthy individuals. *Eur J Cancer* Care 1996, 5, 8-23.
61. Akechi T, Kugaya A, Okamura H, Yamawaki S, Uchitomi Y. Fatigue and its associated factors in ambulatory cancer patients: a preliminary study. *J Pain Symptom Manage* 1999, 17, 42-48.
62. Gijsbers van Wijk CMT, Kolk AM. Sex differences in physical symptoms: the contribution of symptom perception theory. *Soc Sci Med* 1997, 45, 231-246.
63. Smets EMA. Fatigue in cancer patients undergoing radiotherapy. Thesis. University of Amsterdam, The Netherlands, 1997.
64. Trijsburg R, van Knippenberg F, Rijpma S. Effects of psychological treatment on cancer patients: A critical review. *Psychosom Med* 1992, 54, 489-517.

Chapter 6
Quality of life, self-esteem and worries in young adult survivors of childhood cancer

Abstract

Purpose: The aim of the study was to assess quality of life, self-esteem and worries in young adult survivors of childhood cancer and to compare the results with a group of young adults with no history of cancer. The impact of demographic, medical and treatment factors and self-esteem on survivors' quality of life and worries was studied.

Patients and methods: Participants were 400 long-term survivors (LTS) of childhood cancer (age range 16-49 years, 45% female) who had completed treatment an average of 16 years previously and 560 persons (age range 16-53 years, 55% female) with no history of cancer. All participants completed the MOS-24 (Medical Outcome Study Scale), a Worry questionnaire consisting of 3 scales (cancer-specific concerns, general health concerns, present and future concerns), and the Rosenberg Self-Esteem Scale.

Results: Small to moderate differences were found in mean MOS-24 scores between the LTS group and controls (range effect sizes -0.36-0.22). No significant difference was found in the mean self-esteem scores between LTS and controls. Female LTS had more cancer-specific concerns than male LTS. In several related areas of general health, self-image and dying, the LTS group reported less worries than controls, but LTS worried significantly more about their fertility, getting/changing a job and obtaining insurance's. Multiple linear regression analysis revealed that female gender, unemployment, severe late effects/health problems and a low self-esteem were predictors of worse quality of life in survivors. In addition, age at follow-up, unemployment, years since completion of therapy and a low self-esteem were associated with a higher degree of survivors' worries.

Conclusion: Quality of life and the level of self-esteem in LTS of childhood cancer is not different from their peers. Although many LTS worried not more or even less about health issues than their peers, they often are concerned about some present and future concerns. The investigated factors could explain poor quality of life and worries only to a limited extent. Further research exploring determinants and indices of quality of life and worries in LTS is warranted. From a clinical perspective, health care providers can use this knowledge to plan interventions to enhance the quality of life of LTS and to decrease the degree of worries.

Introduction

Over the last decades there has been a dramatic improvement in the survival of children with cancer. Greater than 70 % of children newly diagnosed with acute lymphocytic leukaemia will be in complete continuous remissions 5 years following their initial diagnosis, and the majority of these patients are probably

93

cured of their disease. Survival has also increased for children with solid tumours: approximately 90% of children with Hodgkin's Disease, 92 % of children with Wilms' tumour and nearly 80% of children with non-Hodgkin's lymphoma will be alive 5 years after diagnosis [1]. With the use of combined modalities of therapy, such as surgery, radiation and chemotherapy, centralisation of care and improved supportive care, now over two third of childhood cancer patients are likely to be cured [1].

The clinical success achieved with this patient population has, however, come at some cost in terms of patients' level of functioning and sense of well-being. Both chemotherapy and radiotherapy may have adverse effects upon normal body tissue that may manifest themselves months or even years after completion of therapy. In addition to the development of second malignancies, the areas of greatest concerns are growth and endocrine dysfunction, infertility and serious organ toxicity's [2-5]. Although much has been written about individual toxicity's and late mortality, little is known of the overall morbidity within this population. Stevens and colleagues [6] found in their study, aimed to document the pattern of late effects of childhood cancer treatment, that more than half of the survivors had at least one chronic medical problem. Another study [7] that examined the late effects in young adult survivors showed that approximately 69% of the patients had at least one late effect. Thirty-three percent of patients had a single late effect whereas 36% had two or more late effects.

Although there is growing attention for possible physical disabilities, there is relatively little information about the quality of life (QL) as perceived by young adult survivors. QL is a multidimensional concept which comprises elements of physical, functional, social and psychological health, as well as the patients' perceived health status and well-being [8]. The importance of this definition to childhood cancer survivors is the inclusion of both emotional and social dimensions of health in addition to physical. First of all survivors have to come to terms with the afore-mentioned physical late effects. Furthermore, the life-threatening experience of cancer is in most cases never forgotten. In many ways, survival enhances appreciation for life, while at the same time reminding survivors of their vulnerability. The metaphor of the Damocles syndrome illustrates this dichotomy and how individual survivors interpret this metaphor of life will influence the quality of their survival [9].

To date, most studies have focused on broad adjustment issues such as depression and anxiety, educational achievement, and social functioning. Moreover, these studies have varied in their conclusions. Many studies highlighted several difficulties that some survivors experience in adult life, including problems with schooling and employment, insurability, the ability to marry and have children, and in establishing stable relationships with family and friends [10-17]. In contrast, other authors report more positive findings, concluding that most long-term survivors function well psychologically and do not have significant problems [18-21].

To our knowledge, the QL of young adult childhood cancer survivors is not often investigated with generic measures and compared with the QL of a general population sample. Moe and colleagues [22] and Tebbi and colleagues [23] explored the QL of childhood cancer survivors and a control group by use of the SF-12 and the Rand Health Insurance study General Well-being measure, respectively. Both studies found no statistical differences between the groups with respect to physical health and QL. In the study by Moe and colleagues [22], only the somatisation score on the General Health Questionnaire with items closely related to fatigue demonstrated a significantly higher score for the survivors with acute lymphoblastic leukaemia than for the controls. Apajasalo and colleagues [24] used the 15D (a 15-dimensional questionnaire) to examine the health-related QL of 168 survivors with a range of different malignancies and 129 controls. They found that the QL score of the survivors was significantly better than the controls: survivors reported better levels of vitality, distress, depression, discomfort, elimination and sleeping dimensions. However, their study did not include patients with CNS tumours, due to low survival of these patients. One study [25] used the SF-36 to assess the general QL in childhood bone tumour survivors. Survivors performed significantly poorer than controls with respect to physical functioning and role-functioning-physical.

Self-esteem and self-reported concerns and fears of young adult survivors have been systematically investigated only in a few studies. A positive self-concept is a significant factor influencing overall good mental health and psychological well-being [26,27], and is regarded by major theorists as a basic psychological need [28-30]. The cancer experience is traumatic and the perception of past experiences, especially negative, can be a factor in developing self-esteem [31]. The potential psychosocial problems for survivors such as problems with schooling and employment, insurance denial, difficulties in forming relationships and adverse changes in cosmetic appearance, can result in a possible lack of self-esteem [11]. In several studies childhood cancer survivors reported worries about cancer-specific issues and general health issues even many years beyond the end of their treatment [18,32-35]. It would be useful to predict who is most likely to worry, in order to direct possible resources for psychological help towards those most likely to worry excessively. As noted by Weigers and colleagues [35], understanding the worries of long-term survivors may be important for two reasons. First, degree of worry, or anxiety, has been shown to be an important component of overall adjustment [36]. Second, the self-reported worries of survivors may provide a starting point for the development of effective long-term support for them.

The aim of the present study was to assess quality of life, self-esteem and the degree of worries of young adult survivors of childhood cancer. The results of these survivors were compared with the quality of life, self-esteem and the degree of worries in a comparison group of young adults with no history of cancer. Further, the study examined the relationship between demographic, medical and treatment characteristics, and self-esteem on survivors' quality of life and degree of worries.

Patients and methods

Study group
Data were collected from two samples: young adult survivors of childhood cancer (hereafter referred to as the survivors group) and a reference group of persons with no history of cancer (hereafter referred to as the comparison group).

Survivors group
Patients eligible for participation in the study were those attending the long-term follow-up clinic at The Emma Kinderziekenhuis/Academic Medical Center for their annual evaluation between 1996 until 1999. The long-term follow-up clinic was established in 1996 to monitor long-term sequelae of childhood cancer and its treatment. Patients become eligible for transfer from active-treatment clinics to the follow-up clinic when they completed cancer treatment successfully at least 5 years earlier. Survivors are evaluated annually in the clinic by a paediatric oncologist (persons aged <18 years) or internist-oncologist (persons aged >18 years) for late medical effects, as well as a research nurse or psychologist for psychosocial effects. Study participants had to be aged 16 years or older, have had a pathologic confirmation of malignancy and their cancer had to have been diagnosed before the patients were 19 years of age. Within the study period a total of 443 patients were offered appointments to attend the follow-up clinic. Of these patients, 11 attended but were not included in the study (one person was schizophrenic, 8 were developmentally delayed and 2 were deemed ineligible because of a current health problem causing emotional upset). A further 32 patients did not attend, representing a failed attendance rate over this period of 7%. Survivors who did attend were significantly younger than survivors who did not attend (mean = 24 versus 26 years; P<0.05), but there were no differences with regard to sex or type of diagnosis. A total of 400 patients were approached to take part in this study during their visit at the follow-up clinic and all agreed to participate. After their informed consent the survivors were individually asked to complete a questionnaire by one of the authors (NL). The investigator was present to make sure that the questionnaire was clearly understood. Most survivors had no difficulty in completing the questionnaire, only 2 persons needed some assistance.

Comparison group
Participants were recruited with the help of survivors' general practitioners (GPs). Letters with response cards were send to the GPs of a part of the survivor group (n=540) explaining the purpose of the study and asking for their help in selecting an age-matched control group.
Three hundred and thirty GPs responded (61% response rate), of which 151 stated that they could not participate because of a lack of time: thus, 179 from the notified GP's agreed to take part in the study. These GPs were asked to select 10 persons from their patient's registry lists (starting with a given letter from the alphabet) with a given sex and age range. Those with (a prior history of) cancer were to be excluded. The GPs had to send eligible persons a packet containing the questionnaire, a stamped return address envelope and a cover letter. Two weeks after the original mailing date the GP's had to send another packet with the same content and a reminder letter. Of 1790 questionnaires mailed, 23 were returned because the persons had moved and their address could not be traced. Twenty-four persons let us know that they refused to participate for various reasons (lack of time n=5; not able to understand Dutch n=4; perceived invasion of privacy n=7; other reasons n=8). Of the 1743 remaining questionnaires 1096 completed questionnaires were returned (response rate 63%). Four responses were excluded afterwards because the respondents did not met the inclusion cri-

teria (too young n=3; too old n=1); thus, 1092 responses were included in the final analysis. Half of the persons in the comparison group were asked to fill out the MOS-24 (n=560), whereas another type of QL measure was used for the remaining group. Thus, complete MOS-24 data for this study were obtained from 560 persons.

Measures

Data were collected on sociodemographic characteristics in terms of age at follow-up, gender, marital status (single, living together/married), educational level (low = less than high school, high = high-school or advanced degree), and employment status (unemployed, student/homemaker, employed). In addition, a paediatrician, who had experience regarding the long-term effects of childhood cancer, reviewed the medical record prior to the visit to obtain information about survivor's cancer history and these data were recorded onto structured data coding sheets. Age at diagnosis, time since completion of therapy and duration of treatment were assessed. Diagnosis was categorised into: leukaemia/non-Hodgkin's lymphoma with or without cranial radiation therapy (CRT), solid tumours, and brain/central nervous system tumours (CNS). Treatment was aggregated into three categories: chemotherapy (with or without surgery), radiation therapy (with or without surgery), and combination therapy (chemotherapy and radiation therapy with or without surgery). Finally, late effects and health problems were scored on an adapted version of the Greenberg, Meadows & Kazak's Scale for Medical Limitations [37] by two paediatric oncologists and a paediatric oncology nurse, who were blinded to the survivors' identity. Patients were categorised into the following three groups according to their most serious medical limitation: 1 (mild) = no limitations of activity. This group included children with one kidney and second benign neoplasm's, and no cosmetic or organ dysfunction; 2 (moderate) = no serious restriction of daily life. This group included children with hypoplasia or asymmetry of soft tissue, mild scoliosis and other mild orthopaedic problems, moderate obesity, abnormally short stature, mild hearing loss, cataract, hypothyroidism, delayed sexual maturation, learning delay, enucleation of one eye, small testis, elevated follicle stimulating hormone/luteinising hormone (FSH/LH), alopecia, hypertension, pulmonary diffusion disturbances; 3 (severe) = significant restriction on daily activity or severe cosmetic changes. This group included children with learning delay requiring special education, soft tissue or bone changes that alter appearance, severe asymmetry, absent limb, dental reconstruction, gonadal failure, azoospermia, known sterility, blindness, organ damage that limits activity, second malignant neoplasm's, hemipareses, fatigue that affect daily and social activities. Fifteen percent (n =59) of survivors were categorised as mild, 46% (n = 185) as moderate, and 39% (n =156) as severe.

Respondents' self-assessment of quality of life was measured by the Medical Outcome Study Scale (MOS-24). The MOS-24 is a reliable and valid standardised measure of quality of life and is widely used for medical patients [38-40]. In the original version, six 'dimensional scores' cluster 20 items under physical functioning (extent to which health limits physical activities such as self-care, walking, and climbing stairs), role functioning (extent to which health interferes with usual daily activity such as work, housework, or school), social functioning (extent to which health interferes with normal social activities such as visiting with friends during past month), mental health (general mood or affect, including depression, anxiety, and psychologic well-being during the past month), bodily pain (extent of bodily pain in past 4 weeks), and general health perceptions (overall ratings of current health in general). An energy dimension was added to the questionnaire. Four items, derived from the original RAND battery test [41] were added to the original version. The MOS-24 was scored on a five-point scale. MOS-24 scores were transformed to a scale of 0-100, a higher score indicating a better quality of life state for all scales except pain, in which a high score denotes poor health. The Dutch translation of the MOS-24 proved to be valid and reliable [42,43]. The reliability (i.e. internal consistency) of the MOS-24 in our study, as measured by Cronbach's alpha coefficient, ranged from 0.77 for the vitality scale to 0.86 for the mental health scale.

Self-esteem was assessed with the Rosenberg Self-Esteem Scale (RSE), a self-administered, 10-item, four-point scale with response options ranging from 1 (strongly agree) to 4 (strongly disagree) [44]. The scale measures the self-acceptance aspect of self-esteem or the overall sense of being capable, worthwhile, and competent. Possible scores range from 10 to 40. "Positive" and "negative" items are alternated in an attempt to reduce the effect of respondent set. The scoring direction on five negatively phrased items was reversed, so a higher score now is indicative of higher self-esteem. For the study sample, a sufficient Cronbach's alpha coefficient of 0.89 was calculated of the items.

To investigate the respondents' concerns, we used items from the self-reporting Worry questionnaire by Chesler and colleagues [35]. The questionnaire consisted of 18 items, and issues ranged from general health concerns to concerns about parents. Respondents were asked to indicate the degree of worry on a four-point scale (1 = never worry, 4 = worry a lot). Items composing a subscale were selected after exploring several principal component factor analyses with varimax rotation and inspecting the psychometric features of the items. Inclusion of items on the subscales was based on 1) factor loadings higher than 0.40, 2) no reduction of the Cronbach alpha coefficient of the subscale, and 3) a considerable correlation with the other items of the subscale. When these steps were completed, three reliable subscales remained: a) cancer-specific concerns (5 items), b) general health concerns (4 items), and 3) present and future concerns (9 items). The items in each subscale are listed in Figure 1. As the cancer-specific concerns addressed issues of former illness and treatment, only survivors were asked to complete this subscale. The other two subscales were presented to both survivors and controls.

Figure 1. Worry questionnaire

Cancer-specific concerns (Cronbach alpha 0.80)
1. Having a relapse
2. My children getting cancer
3. More complicated future treatments
4. Having another cancer when I am older
5. The effects of my illness on my sisters/brothers

General health concerns (Cronbach alpha 0.64)
1. Getting a cold or the flu
2. If I am as healthy as other people my age
3. Having headaches
4. Getting tired

Present and future concerns (Cronbach alpha 0.78)
1. Dying
2. How my body looks
3. Doing well in school (or at work)
4. My parents' health
5. Losing friends
6. Obtaining life or medical insurance
7. Getting or changing a job
8. My parents' finances
9. Whether I can have children

Statistical analysis

The Statistical Package for Social Sciences (SPSS) Windows version 9.01 was used for all statistical analyses. Descriptive statistics were performed for all of the variables. Differences in demographic characteristics between survivors and the comparison group were analysed with Chi-square tests or Student's t-tests. Differences between the mean scores of the MOS-24, the RSE and the Worry scales in the survivors and the comparison group were tested with the Student's t-test. Differences between the mean scores were analysed for sex and age groups ($\leq$ 25 years and $\geq$ 26 years). To examine the magnitude of these differences, effect sizes ($= d$) were calculated by dividing the difference between a given mean score of the survivor group and the mean score in the comparison group by the standard deviation of scores in the comparison group. We considered effect sizes up to 0.20 to be small, effect sizes of about 0.50 to be moderate, and effect sizes of about 0.80 to be large [45].

Univariate relationships between demographic, medical and treatment characteristics and self-esteem on the one hand, and by MOS-24 and Worry scores on the other hand were assessed by Pearsons correlation coefficients, Student's t-test or one way ANOVA. To investigate which variables predict survivors quality of life and concerns, all significant characteristics (set at $P \leq 0.20$) identified from univariate analysis were studied with multiple regression analysis (with a stepwise forward selection strategy). For three variables (employment status, diagnosis, and treatment) we created dummy variables and took the first category (unemployment, leukaemia/non-Hodgkin's lymphoma without CRT, and chemotherapy with or without surgery) as reference for the analysis. For each regression, the explained variance (R square) was estimated.

Results

Characteristics of survivors and comparison group
Information about the demographic, medical and treatment characteristics of the survivor and the comparison group is listed in Table 1. The survivors' age at diagnosis ranged from 0 to 19 years (median 8), the range of the time since completion of therapy was 5 to 33 years (median 15). The median duration of treatment was 10 months (range 0-170 months). The survivors were treated for a variety of cancers. The most frequent diagnoses were leukaemia, non-Hodgkin's lymphoma, Hodgkin's disease, Wilms' tumour and brain/CNS tumours. We compared the distribution of cancer diagnoses of our study population with the distribution of diagnoses of the survivors, who were known to be alive but were not yet seen in the follow-up clinic. We found an over-representation of leukaemias and lymphomas in our study group. From logistic considerations many survivors with these diagnoses were seen in the follow-up clinic during the first years after the clinic was set up.

The median age of the comparison group was 26 years (range 16-53 years), this was slightly higher than that of the survivors (median 23; range 16-49 years). The differences in proportions of men and women in the groups were significant, there were more men than women in the survivor group (55% versus 45%), whereas the reverse was true for the comparison group (45% versus 55%). Further, a Chi-square analysis showed that survivors and controls differed in terms of marital status, more survivors were single compared to controls. Because these could be typical features of the survivor group, univariate analysis of variance was performed to examine the influence of age at follow-up, sex, subject status, marital status, and the interaction of these variables on the different subscales of the MOS-24 and the Worry questionnaire. We found no significant effect of subject status and marital status on any of the dependent variables. Therefore, the influence of marital status on the dependent variables was not further accounted for.

Table 1. Demographic, medical and treatment characteristics of the study group

Variable	Survivors (n=400)		Comparison group (n=560)	P-value
Age at follow-up (years) Mean ± SD	24 ± 4.9		26 ± 5.3	< 0.001[*]
Sex % Men Women	55 45		45 55	< 0.05[#]
Marital status % Single Married/cohabitant	72 28		44 56	< 0.001[#]
Educational level % Lower level Higher level	63 37		58 42	0.11[#]
Age at diagnosis (years) Mean ± SD	8 ± 4.6			
Time since completion of therapy (years) Mean ± SD	16 ± 5.6			
Duration of treatment (months) Mean ± SD	16 ± 21.6			
Diagnosis Leukaemia/non-Hodgkin's lymphoma without CRT Leukaemia/non-Hodgkin's lymphoma with CRT Solid tumour Brain/CNS tumour	N 104 89 164 43	% 26 22 41 11		
Treatment Chemotherapy (with or without surgery) Radiation therapy (with or without surgery) Combinationtherapy (chemotherapy and radiation therapy with or without surgery)	 168 36 196	 42 9 49		
Medical limitations % None/mild Moderate Severe	N 59 185 156	% 15 46 39		

[*] t-test
[#] chi-square

Table 2. Mean (SD) MOS-24 and RSE scores for the survivors and the comparison group in relation to group, sex and age at follow-up

Scale	Survivors (n=400)	controls (n=560)	Males survivors (n=219)	Males controls (n=253)	Females survivors (n=181)	Females controls (n=307)	≤ 25 years survivors (n=272)	≤ 25 years controls (n=258)	≥ 26 years survivors (n=128)	≥ 26 years controls (n=302)
Physical functioning[a]	87.7 (23.8)#	90.8 (19.8)#	91.3 (21.1)	93.8 (15.3)	83.2 (26.0)#	88.4 (22.6)#	88.3 (23.2)#	92.0 (18.0)#	86.5 (25.0)	89.8 (21.3)
Role functioning[a]	90.8 (26.3)	91.2 (26.5)	93.2 (22.9)	96.3 (17.1)	87.9 (29.7)	87.0 (31.7)	93.8 (22.3)	91.7 (25.7)	84.4 (32.5)#	90.7 (27.3)#
Social functioning[a]	91.5 (19.6)	91.9 (15.1)	94.1 (16.2)	95.4 (10.1)	88.3 (22.6)	89.0 (17.7)	92.4 (17.6)	92.4 (14.8)	89.4 (23.1)	91.5 (15.4)
Mental health[a]	79.0 (16.5)	78.5 (14.7)	81.2 (15.0)	81.9 (13.1)	76.3 (17.8)	75.9 (15.4)	79.4 (15.9)	78.5 (15.1)	78.0 (17.6)	78.6 (14.3)
Vitality[a]	71.5 (19.7)#	68.1 (16.6)#	75.6 (18.3)#	71.9 (15.0)#	66.7 (20.3)	65.2 (17.3)	72.5 (19.6)#	69.3 (16.9)#	69.6 (19.9)	67.1 (16.3)
Bodily pain[b]	14.6 (24.7)*	23.6 (19.7)*	8.8 (19.9)*	17.9 (22.5)*	21.5 (28.1)#	28.3 (26.4)#	13.8 (23.9)*	23.9 (25.1)*	16.2 (26.4)#	23.3 (25.4)#
General health perception[a]	80.7 (19.4)*	76.6 (18.4)*	83.4 (17.2)#	79.8 (15.4)#	77.5 (21.4)	74.0 (20.2)	81.7 (17.9)#	77.7 (18.6)#	78.6 (22.2)	75.7 (18.2)
Self-esteem	32.1 (4.9)	32.3 (4.6)	33.4 (4.5)	33.6 (4.2)	30.6 (5.0)	31.2 (4.7)	32.1 (5.1)	32.4 (4.8)	32.2 (4.8)	32.3 (4.5)

[a] High scores = good health
[b] High scores = poor health
Statistically significant difference (p <0.05) in means by t-tests
* Statistically significant difference (p <0.001) in means by t-tests

Quality of life in survivors versus comparison group
In Table 2 the mean scores and standard deviations for the different subscales of the MOS-24 are presented for the survivors and the comparison group in relation to group, sex and age at follow-up. The group of survivors scored significantly lower levels of physical functioning (P 0.03, effect size = d = -0.16), but statistically higher on vitality (P 0.005, d= 0.21) and general health perceptions (P 0.001, d= 0.22) than controls. Further, survivors reported significantly less bodily pain than their peers (P 0.001, d= -0.36). In general, females perceived their quality of life worse than their male peers on all quality of life dimensions. Female survivors reported significantly lower levels of physical functioning than female controls (d= -0.23). In particular, they were more likely to be limited in performing strenuous physical activities (e.g. running, lifting heavy objects, participating in strenuous sports and climbing stairs). Additionally, male survivors reported significantly higher levels of vitality (d= 0.25), less pain (d= -0.40) and better general health perceptions (d= 0.23) than males in the comparison group. Within the age groups, survivors 25 years old or younger reported significantly more limitations in their physical functioning (d= -0.21), but statistically less bodily pain (d= -0.40) than their peers. These survivors also had significantly higher levels of vitality (d= 0.19) and better general health perceptions (d= 0.22) than controls. Survivors 26 years and older reported significantly more limitations in their role functioning (d= -0.23) and had significantly less bodily pain (d= -0.28) than their peers.

Self-esteem in survivors versus comparison group
We found no difference in the mean self-esteem score between the survivors and controls as a group. When we looked to the groups in relation to sex and age at follow-up (Table 2), we only found that the mean scores of women were significantly lower than the mean scores of men (P <0.001, d= -0.58).

Worries in survivors versus comparison group
As is shown in Table 3, female survivors reported more cancer-specific concerns than male survivors (P 0.003, d= 0.31). Many survivors (54%) worried about having a relapse, and 43% expressed their concerns about the health of their future children. In addition, 50% of the survivors reported that they worried about having another cancer when they are older. We found no significant differences in mean general health concerns between survivors and controls in relation to sex and age groups. With respect to present and future concerns, female survivors reported worrying significantly less than their female peers (d = -0.25). Survivors 26 years or older had also less present and future concerns than controls (d = -0.38). In general, women reported worrying significantly more than men on all subscales.

Table 3. Mean (SD) Worry scores for the survivors and the comparison group in relation to group, sex and age at follow-up

Scale	survivors (n=400)	controls (n=560)	Males survivors (n=219)	controls (n=253)	Females survivors (n=181)	controls (n=308)	≤ 25 years survivors (n=272)	controls (n=258)	≥ 26 years survivors (n=128)	controls (n=302)
Cancer-specific concerns[a][b]	7.7 (2.9)	-	7.3 (2.7)*	-	8.2 (3.0)*	-	7.78 (3.0)	-	7.5 (2.6)	-
General health concerns[c]	5.5 (2.0)	5.7 (1.9)	5.2 (1.8)	5.2 (1.6)	5.9 (2.9)	6.0 (1.9)	5.46 (1.9)	5.74 (1.9)	5.6 (2.2)	5.6 (3.8)
Present and future concerns[d]	14.7 (4.6)	15.3 (4.0)	14.3 (4.2)	13.9 (3.2)	15.3 (5.0)*	16.3 (4.3)*	15.14 (4.9)	15.81 (4.4)	13.9 (3.8)*	14.7 (3.5)*

a these questions were not administered to the persons in the comparison group
b scores range from 5 to 20, c scores range from 4 to 16, d scores range from 9 to 36 (higher scores indicates more concerns)
* statistically significant difference (p < 0.05) in means by t-tests

Table 4. Association (expressed in correlation coefficient) between survivors' demographic and medical characteristics and self-esteem on the one hand and quality of life and worries on the other hand

Characteristics	PF	RF	SF	BP	MH	VT	GH	C-SC	GHC	PFC
Demographic and medical characteristics										
age at follow-up (years)	-0.07	-0.14*			-0.07		-0.08			0.11*
age at diagnosis (years)						0.07				-0.09
duration of treatment (months)				0.07	-0.08	-0.10				
time since completion of therapy (years)		-0.09	-0.07		-0.11*	-0.11*	-0.08	-0.08		
medical limitations	-0.28**	-0.23**	-0.22**	0.12*	-0.16*	-0.24**	-0.27**		0.16*	0.07
Self-esteem	0.20**	0.16*	0.30**	-0.26**	0.58**	0.49**	0.43**	-0.25**	-0.40**	-0.42*

Abbreviations: PF: physical functioning, RF: role functioning, SF: social functioning, BP: bodily pain, MH: mental health, VT: vitality, GH: general health perceptions, C-SC: cancer-specific concerns, GHC: general health concerns, PFC: present and future concerns
NOTE. Only characteristics significant at the p ≤ 0.20 level are shown.
 * p < 0.05
 ** p < 0.001

When we looked at the concerns on item level, survivors reported a significantly higher degree of worry than the comparison group about whether they were as healthy as their peers (P 0.004), their fertility (P <0.001), getting or changing a job (P 0.002), and obtaining life or medical insurance (P <0.001). In contrast, survivors reported worrying significantly less about getting a cold or the flu (P 0.01), dying (P <0.001), having headaches (P 0.002), how their body looked (P 0.001), their parents' health (P <0.001), and losing friends (P <0.001).

Association between survivors' demographic, medical and treatment characteristics and self-esteem and quality of life and worries

Univariate associations between survivors' demographic, medical and treatment characteristics and self-esteem on the one hand and MOS-24 and Worry scores on the other hand are shown in Tables 4 and 5.

After introduction of the most important univariate associations (P ≤0.20) in multivariate regression analysis, many of the observed relationships disappeared (Table 6). However, unemployment and lower self-esteem remained mostly associated with poorer quality of life scores and a higher degree of concerns. Female gender explained lower physical functioning, bodily pain and lower vitality. Higher age at follow-up remained only related to a higher degree of present and future concerns. Lower educational level showed an association with impaired physical functioning and bodily pain. With respect to medical and treatment characteristics, fewer years since completion of therapy was associated with more cancer-specific concerns. Severe late effects/health problems were associated with worse quality of life scores on five subdimensions. Finally, treatment (combination therapy with or without surgery) explained impaired levels of mental health and vitality. The selected characteristics explained only a small proportion of the variability (R^2) of the MOS-24 and Worry scores: 8% - 37%.

Discussion

As an increasing amount of childhood cancer survivors enter adulthood, questions arise regarding their quality of life. Do they differ from their peers with regard to QL? Do they have more or other fears and concerns than their peers? And how is their self-esteem doing, is that affected by the cancer experience? In the current study the QL, the level of self-esteem and the degree of worries in a group of young adult survivors of childhood cancer was compared with a sample of young adults with no history of cancer. In addition, the relationship between demographic, medical and treatment factors and self-esteem and survivors' QL and degree of worries was examined. The following discussion summarises the main findings, considers the clinical implications, and identifies several directions for future research.

Table 5. Survivors mean MOS-24 and Worry scores by demographic, medical and treatment characteristics

Characteristics	PF	RF	SF	BP	MH	VT	GH	C-CS	GHC	PFC
Sex										
Men	91.3*	93.2*	94.1*	8.8**	81.2*	75.6**	83.4*	7.3*	5.2*	14.3*
Women	83.2	87.9	88.3	21.5	76.3	66.7	77.5	8.2	5.9	15.3
Marital status										
Single		92.0		13.5						14.9
Married/cohabitant		87.7		17.3						14.3
Educational level										
Low	85.1*		89.4*	17.4*		70.2*	78.9*	7.9*	5.7*	14.4*
High	92.1		94.9	9.8		73.9	83.9	7.3	5.2	15.0
Employment status										
Unemployed	69.4**	70.2**	81.4*	26.7*	73.2*	61.2*	66.7**		6.2*	16.2*
Student/homemaker	88.9	93.4	91.6	12.6	77.6	71.4	80.5		5.6	15.2
Employed	90.3	93.0	93.1	13.4	80.7	73.5	83.4		5.3	14.4
Diagnosis										
Leukaemia/NHL without CRT			93.9							
Leukaemia/NHL with CRT			88.8							
Solid tumours			92.6							
Brain/CNS tumours			87.0							
Treatment										
CT (with or without surgery)		94.1*	93.9*	11.5	79.1	72.4	82.9			
RT (with or without surgery)		77.8	82.2	18.1	73.1	65.7	74.7			
Combination therapy (CT+RT with or without surgery)		90.3	91.0	16.5	80.0	71.9	80.0			

Abbreviations: PF: physical functioning, RF: role functioning, SF: social functioning, BP: bodily pain, MH: mental health, VT: vitality, GH: general health perceptions, C-SC: cancer-specific concerns, GHC: general health concerns, PFC: present and future concerns, NHL: non-Hodgkin's lymphoma, CRT: cranial radiation therapy, CNS: central nervous system, CT: chemotherapy, RT: radiation therapy

NOTE. Only characteristics significant at the $p \leq 0.20$ level are shown.

* $p < 0.05$

** $p < 0.00$

Table 6. Forward stepwise linear regression models (Beta)[a] to explain survivors' quality of life dimensions and worries

Characteristics	PF	RF	SF	BP	MH	VT	GH	C-SC	GHC	PFC
Demographic characteristics										
Sexe (female)	-0.15*					-0.11*				-0.14*
Age at follow-up (years)	0.11*									
Educational level (higher level)				-0.10*						-0.14
Employment status[b]										
Student/homemaker	0.32**	0.34**	0.17*	-0.23*	0.04	0.18*	0.25*		-0.12	
Employed	0.40**	0.39**	0.24*	-0.27**	0.15*	0.25*	0.35**		-0.18*	-0.19*
Medical and treatment characteristics										
Years since completion of therapy								-0.10*		
Medical limitations	-0.23**	-0.17*	-0.15*			-0.15*	-0.18**			
Treatment[c]										
Radiation therapy (with or without surgery)					0.01	0.03				
Combination therapy (with or without surgery)					0.13*	0.12*				
Self-esteem			0.23**	-0.17*	0.56**	0.42**	0.36**	-0.22**	-0.35**	-0.42**
Total R²[a]	19%	14%	16%	15%	37%	31%	28%	8%	18%	20%

Abbreviations: PF: physical functioning, RF: role functioning, SF: social functioning, BP: bodily pain, MH: mental health, VT: vitality, GH: general health perceptions, C-SC: cancer-specific concerns, GHC: general health concerns, PFC: present and future concerns

[a] R^2 is the percentage of the total variation of the dependent variable score that is explained by the independent variables together

[b] reference group = unemployment, [c] reference group = chemotherapy (with or without surgery)

* $p < 0.05$

** $p < 0.001$

To our knowledge, the current study is the first to evaluate QL, self-esteem and worries in a large sample of young adult survivors of childhood cancer compared with a control group. The response rate among the comparison group was highly satisfactory for a mailed survey. Although the survivors and comparison group differed with regard to marital status, it is doubtful that this difference was influential because QL and worries were not significantly related to marital status. Nonetheless, several limitations existed. First, as with all survey studies, the likelihood of respondents being those particularly interested in the topic is high. Second, the respondents may be those who suffered from physical problems and were motivated to respond. Third, the findings of this study are limited by the heterogeneous patient groups of cancer diagnoses, treatment regimens and the crude assessment categories used.

With regard to QL, we only found small differences in mean MOS-24 scores between survivors as a group and their controls. Survivors reported lower levels of physical functioning, but they had higher levels of vitality, less bodily pain and their general health perceptions were better than controls. However, the effect sizes of the significant differences were all of a small and moderate magnitude. When we looked to the groups in relation to sex and age at follow-up, we found that women perceived their QL worse than their male peers on all dimensions. Further, we found some differences between male survivors and male controls with respect to vitality, bodily pain, and general health perception and between female survivors and female controls with respect to physical functioning and bodily pain. Survivors and controls aged 25 years old or younger differed with respect to physical functioning, vitality, bodily pain and general health perception. Finally, survivors aged 26 years or older reported more limitations in their role functioning than their peers, but had less bodily pain than their controls. But also here the effects sizes were small. The findings suggest that the QL of young adult survivors of childhood cancer, as measured with a generic instrument, is more or less as good, and in some dimensions even better as that of their peers. Because no other studies where one measured the QL with the MOS-24 are available, it is difficult to compare our results with other findings. As described in the Introduction, only one study used the SF-36, which is a revised, 36-item version of the MOS instrument [25], to assess the long-term effects on QL after rotationplasty in a group childhood bone tumour survivors. In line with our results, the authors found that survivors performed significantly poorer with respect to physical functioning, but this is not surprising because their sample consisted of bone tumour survivors who underwent a rotationplasty. Also part in line with our results, they found that survivors' mean scores on most other dimensions were, although not statistically significant, generally higher (indicating a better QL) than those of their healthy counterparts. In the Apajasalo et al. study [24], survivors reported also significantly better levels (less problems) on several QL dimensions of the 15-D questionnaire.

It is commonly assumed that childhood cancer and its treatment have a severe, negative impact on the QL of patients. However, there are several studies, in which self-report questionnaires are used, showing that childhood cancer survivors are not more anxious [22,46-48], more depressed [18,22,47,49,50], or less well emotional adjusted [18,22,23,46-48,51] than siblings or the general, healthy population. In some cases, this seems to be in contrast to the everyday experience of physicians, nurses and other health care professionals, and to the results of studies in which a more negative picture is given [16,52-55]. Our findings in the childhood cancer survivor literature are by no means unique. When one scans the adult cancer literature and compares cancer patients with other (non-cancer) patients, one often will find no differences in most of the indicators of life quality, such as overall QL, mental functioning, fear, depression, well-being, daily activities and resumption of one's work [56,57]. It is possible that the instrument we used was unsuitable for measuring health-related quality of life in survivors. Although generic instruments measure health functioning across a wide variety of chronic illnesses and cancer and provide a common data base for comparing results [58], they may not identify issues unique to the cancer experience and do not contain a longer view of cancer survivorship with specific concerns and needs long term [59]. The development of a specific instrument for childhood cancer survivors may overcome this problem. It is also possible that cancer and its treatment influence QL less than is generally expected. Patterson [60] introduced the term 'resilience', which describes the strengths and abilities of patient families who can 'bounce back' from the stress and challenges they face, and eliminate, or minimise negative outcomes. The lack of differences between the patients and the controls may also be explained by the theory of the so-called 'response-shift'. Response shift refers to the idea that as a result of health state changes, a person may undergo changes in internal standards, values or conceptualisations [56,61,62]. In our setting it could imply that living with the physical and/or psychological consequences of cancer and its treatment for a relatively long period could have changed survivors' standard of measurement concerning QL and as a result QL has been overreported. Previous research has documented that response shift may adversely affect the results of self-reported outcomes in clinical trials and other longitudinal studies [63].

Other explanations of the lack of differences between patients and controls are given in the paediatric oncology literature. According to Apajasalo and colleagues [24], the observed excellent health-related QL could be explained in two ways. First, survivors may, after experiencing a life-threatening disease, find their present life more satisfying and possible defects in their present health status less significant. Second, denial is likely to be involved. In the authors' opinion, fear of recurrence, employment and insurance problems as well as potential discrimination by peers and partners may lead to a situation where a person ignores even objective symptoms and findings. These denial mechanisms may compensate or even overcompensate the objectively measurable late effects. In another study where adult survivors of childhood cancer were found to be at least as well adjusted as their peers, the researchers postulated "that the apparent mental health of the survivors is defensive-a kind of bravado coping style masking underlying suffering and maladjustment" [18]. Further research in this field, and especially assessing response shift is of utmost importance to further QL research. Until a solution to the above problems has been found, we agree with Breetveld and van Dam, who recommend that answers in questionnaires concerned with QL, psychological distress and the like should be approached with due caution [56].

Also the level of self-esteem and the degree of worries did not differ much between the survivors and the comparison group. The results with respect to self-esteem is consisted with previous reports [11,18,46]. The findings about the concerns of survivors are more or less in line with the Weigers and colleagues {35]. They also found that in several related areas of general health, self-image and dying, the survivors worried less than the comparison sample did. Given that a side effect of some cancer treatment is infertility, it is not surprising that many survivors in our study reported worrying significantly more about their fertility than their peers. Worries about reproductive capacity were reported in other studies as well [18,34,35]. These studies found also that male survivors reported less often concerns about fertility too. In our study, however, we found that the majority of our male survivors worried about fertility too. This is not surprising because cancer treatment can have a negative impact on the fertility of males and females alike. The finding that survivors worried significantly more about getting or changing a job and obtaining life or medical insurance may be consistent with the possible risk of job and insurance discrimination faced by some childhood cancer survivors.

When we examined the relationship between demographic, medical and treatment characteristics and self-esteem and QL and degree of worries, we found that, among the demographic factors, female gender was significantly associated with poorer physical functioning, bodily pain, and vitality. This could be due to the fact that women are simply more inclined to actually report the symptoms they experience than men, both in health surveys and in the consultation room [64]. Raised with the "boys don't cry" doctrine, men might be more reserved about their health problems, whereas women might show a greater willingness to discuss symptoms with others [65]. This seems to be in line with the finding in our study that men perceived their QL as better, has higher mean self-esteem scores, and worried less on all Worry scales than women.

Older age at follow-up was found to be a predictor of present and future concerns, which makes sense instinctively. It is not difficult to imagine that how older the survivor will become, the more likely one is to worry about such things as dying, getting children, and so on.

Unemployment was found to be associated with a lower level of QL on all dimensions and with a higher degree of general health concerns and present and future concerns. It should be noted, however, that the unemployed group was small (n = 42) and more than half of these survivors were, partly or fully, officially declared unfit for work because of medical problems. It makes sense that these survivors report lower levels of QL and a higher number of worries.

Fewer years since completion of therapy was associated with more cancer-specific concerns. Concerns about relapse are an almost universal response in the first months and years after active treatment ends. When times passes by and all goes well, these fears will normally fade a little into the background, bothering the survivor less often. These remarks may be true for the other cancer-specific worries as well.

It is not surprising that survivors with severe late effects/health problems had a worse QL than survivors with moderate and none/mild limitations, especially since survivors with severe medical limitations are more likely to have symptoms related to their disease. We noticed that there was no relation between severe late effects/health problems and worries. One should expect that the more severe the medical problems, the higher the degree of worries would be, but that seems not be the case.

Self-esteem was found to have an association with most of the QL dimensions and worries. This suggest that lower levels of self-esteem are associated with lower levels of QL and with more worries. Instantly the possible confounding between self-esteem and the outcome measures should be considered. Whether self-esteem give rise to a worse QL and a higher degree of worries, or a worse QL and a higher degree of worries lead to a lower self-esteem is not possible to answer. The finding that self-esteem is strongly associated with QL is consistent with prior research findings [66,67]. It also lend support to the study by Dirksen [68], in which self-esteem was found to be an important determinant of subjective well-being. These results underscore the importance of assessing and supporting self-esteem in the long-term survivor of cancer as means of enhancing QL. It may be that those survivors with cancer who have a low-self-esteem, are in need of more help and support. This assessment and support could begin with the survivor in the early stages of survivorship and continue into long-term survival. Specific strategies and approaches to increasing self-esteem, such as support groups or coping skills training, need to be explored and evaluated.

In conclusion, the present findings indicate that the QL and level of self-esteem in young adult survivors of childhood cancer is not different from their peers. Although many long-term survivors worried not more or even less about some issues than their peers, they often are concerned about their own health and some present and future concerns, such as fertility, getting a job and obtaining insurance's. Female gender, age at follow-up, unemployment, years since completion of therapy, severe late effects/health problems and self-esteem could explain variations in QL and worries only to a limited extent. Therefore, other potential determinants of QL and worries should be explored. In the mean time, these findings should be kept in mind by health care providers working with childhood cancer survivors. They can use this knowledge to plan interventions to enhance the QL and decrease the degree of worries. In addition, health indices other than the MOS-24 should be examined. Perhaps we should focus more on the long-term impact of different treatment on QL [69]. It is important to document how varying therapeutic modalities may give rise to different long-term effects. Such information can establish if there are any residual effects of one treatment but not another and if there are treatment-related decrements in QL that vary in the short term and long term. Further, little is known about the impact of persistent effect of cancer treatment on survivors' QL. Survivors may learn to live with and adjust to their possible limitations, they may continue to experience problems to the same degree as during shortly after completion of therapy, or they may have decreased tolerance of disability with the passage of time (i.e. an enhanced QL, an unchanged QL, or a worsened QL, respectively) [69]. It is also important to identify the subgroups of survivors who have problems in stead of evaluating only differences between survivors as a whole and their controls. Although many young adult survivors function as well as the general population, there will be survivors who may continue to experience problems that can affect their QL many years after treatment. Given the range of outcome, follow-up of survivors is considered essential [4,7,70]. Not only investigation of the risk factors and causes of all possible adverse outcomes of childhood cancer and its treatment is important, early identification of late effects and appropriate intervention may be in the interest of the individual survivor. It is possible that some problems can be prevented, others remediated and survivors' QL may be enhanced with some provision of appropriate care. Therefore, counselling and advice should be available to facilitate adaptation to such problems and limitations.

Acknowledgement
We are grateful to the people willing to participate in our study.

References

1. Smith MA, Gloeckler Ries LA. Childhood cancer: incidence, survival, and mortality. In: Pizzo PA, Poplack D, eds. Principles and Practice of Pediatric Oncology. Philadelphia: Lippincott Williams & Wilkins 2002, 1-12.
2. Hawkins MM, Draper GJ, Kingston JE. Incidence of second primary tumours among childhood cancer survivors. *Br J Cancer* 1987, 56, 339-347.
3. Nicholson HS, Mulvihill JJ, Byrne J. Late effects of therapy in adult survivors of osteosarcoma and Ewing's sarcoma. *Med Pediatr Oncol* 1992, 20, 6-12.
4. Hawkins MM, Stevens MC. The long-term survivors. *Br Med Bull* 1996, 52, 898-923.
5. Sklar CA, Constine LS. Chronic neuroendocrinological sequelae of radiation therapy. *Int J Radiation Oncol Biol Physics* 1995, 31, 1113-1121.
6. Stevens MC, Mahler H, Parkes S. The health status of adult survivors of cancer in childhood. *Eur J Cancer* 1998, 34, 694-698.
7. Oeffinger KC, Eshelman DA, Tomlinson GE, Buchanan GR, Foster BM. Grading of late effects in young adult survivors of childhood cancer followed in an ambulatory adult setting. *Cancer* 2000, 88, 1687-1695.
8. de Haan RJ, Aaronson N, Limburg M, Langton Hewer R, van Crevel H. Mesuring quality of life in stroke. *Stroke* 1993, 24, 320-327.
9. Leigh SA, Stovall EL. Cancer Survivorship. Quality of Life. In: King CR, Hinds PS, eds. Quality of Life. From Nursing and Patient Perspectives. Sudbury, Massachusetts: Jones and Bartlett Publishers 1998, 287-300.
10. Byrne J, Fears TR, Steinhorn SC et al. Marriage and divorce after childhood and adolescent cancer. *JAMA* 1989, 262, 2693-2699.
11. Evans SE, Radford M. Current lifestyle of young adults treated for cancer in childhood. *Arch Dis Child* 1995, 72, 423-426.
12. Haupt R, Fears TR, Robison LL et al. Educational attainment in long-term survivors of childhood acute lymphoblastic leukemia. *JAMA* 1994, 272, 1427-1432.
13. Hays DM, Landsverk J, Sallan SE, Hewett KD, Patenaude AF, Schoonover D, Zilber SL, Ruccione K, Siegel SE. Educational, occupational, and insurance status of childhood cancer survivors in their fourth and fifth decades of life. *J Clin Oncol* 1992, 10, 1397-1406.
14. Kelaghan J, Myers MH, Mulvihill JJ, Byrne J, Connelly RR, Austin DF, Strong LC, Meigs JW, Latourette HB, Holmes GF. Educational achievement of long-term survivors of childhood and adolescent cancer. *Med Pediatr Oncol* 1988, 16, 320-326.
15. Rauck AM, Green DM, Yasui Y, Mertens A, Robison LL. Marriage in the survivors of childhood cancer: a preliminary description from the Childhood Cancer Survivor Study. *Med Pediatr Oncol* 1999, 33, 60-63.
16. Zeltzer LK. Cancer in adolescents and young adults psychosocial aspects. Long-term survivors. *Cancer* 1993, 71, 3463-3468.
17. Langeveld NE, Ubbink MC, Last BF, Grootenhuis MA, Voute PA, de Haan RJ. Educational achievement, employment and living situation in long-term young adult survivors of childhood cancer in the Netherlands. *Psychooncology* 2002, 11, 1-13.
18. Gray RE, Doan BD, Shermer P et al. Psychologic adaptation of survivors of childhood cancer. *Cancer* 1992, 70, 2713-2721.
19. Green DM, Zevon MA, Hall B. Achievement of life goals by adult survivors of modern treatment for childhood cancer. *Cancer* 1991, 67, 206-213.
20. Kazak AE. Implications of survival: Pediatric Oncology Patients and their Families. In: Bearison A, Mulhern R, eds. Pediatric Psychooncology. New York: Oxford University Press 1994, 171-192.
21. Meadows AT, McKee L, Kazak AE. Psychosocial status of young adult survivors of childhood cancer: a survey. *Med Pediatr Oncol* 1989, 17, 466-470.
22. Moe PJ, Holen A, Glomstein A, Madsen B, Hellebostad M, Stokland T, Wefring KW, Steen-Johnsen J, Nielsen B, Borsting S, Hapnes C. Long-term survival and quality of life in patients treated with a national all protocol 15-20 years earlier: IDM/HDM and late effects? *Ped Hemat Oncol* 1997, 14, 513-524.
23. Tebbi CK, Bromberg C, Piedmonte M. Long-term vocational adjustment of cancer patients diagnosed during adolescence. *Cancer* 1989, 63, 213-218.

24. Apajasalo M, Sintonen H, Siimes MA, Hovi L, Holmberg C, Boyd H, Makela A, Rautonen J. Health-related quality of life of adults surviving malignancies in childhood. *Eur J Cancer* 1996, 32A, 1354-1358.
25. Veenstra KM, Sprangers MA, van der Eyken JW, Taminiau AH. Quality of life in survivors with a Van Ness-Borggreve rotationplasty after bone tumour resection. *J Surg Oncol* 2000, 73, 192-197.
26. Coopersmith S. The antecedents of self-esteem. Palo Alto, CA: Consulting Psychologists Press 1981.
27. Shavelson RJ, Bolus R. Self-concept: The interplay of theory and methods. *J Educational Psychol* 1982, 74, 3-17.
28. Erikson EH. Childhood and society. New York: W.W.Norton 1950.
29. Freud A. Normality and pathology in childhood: Assessments of development. New York: International Universities Press 1965.
30. Kohut H. The analysis of the self. New York: International Universities Press 1971.
31. Stanwyck DJ. Self-esteem through the life span. *Family Community Health* 1983, 6, 11-28.
32. Haase JE, Rostad M. Experiences of completing cancer therapy: children's perspectives. *Oncol Nurs Forum* 1994, 21, 1483-1494.
33. Smith K, Ostroff J, Tan C, Lesko L. Alterations in self-perceptions among adolescent cancer survivors. *Cancer Investigation* 1991, 9, 581-588.
34. Wasserman AL, Thompson EI, Wilimas JA, Fairclough DL. The psychological status of survivors of childhood/adolescent Hodgkin's disease. *AJDC* 1987, 141, 626-631.
35. Weigers ME, Chesler MA, Zebrack BJ, Goldman S. Self-reported worries among long-term survivors of childhood cancer and their peers. *J Psychosoc Oncol* 1998, 16, 1-23.
36. Holland J, Rowland J. Handbook of Psychooncology. New York: Oxford University Press 1990.
37. Greenberg HS, Kazak AE, Meadows AT. Psychologic functioning in 8- to 16-year-old cancer survivors and their parents. *J Pediatr* 1989, 114, 488-493.
38. Stewart AL, Hays RD, Ware JE. The MOS Short-form General Health Survey. Reliability and validity in a patient population. *Med Care* 1988, 26, 724-735.
39. Stewart AL, Greenfield S, Hays RD, Wells K, Rogers WH, Berry SD, McGlynn EA, Ware JE. Functional status and well-being of patients with chronic condition: Results from the Medical Outcome Study. *JAMA* 1989, 262, 907-913.
40. Wells KB, Stewart AL, Hays RD, Burnam A, Rogers WH, Daniels M, Berry S, Greenfield S, Ware JE. The functioning and well-being of depressed patients. Results from the Medical Outcomes Study. *JAMA* 1989, 262, 914-919.
41. Stewart AL, Ware JE. Mesuring functioning and well-being. Durham: Duke University Press 1992.
42. Kempen GIJM. The measurment of the health status of the elderly. A Dutch version of the MOS. *Tijdschr Gerontol Geriatr* 1992, 23, 132-140.
43. Kempen GIJM. The MOS short-form general health survey: single item versus multiple measures of health-related quality of life: some nuances. *Psychol Rep* 1992, 70, 608-610.
44. Rosenberg M. Society and the adolescent self-image. Princeton, NJ: Princeton University Press 1965.
45. Cohen J. Statistical power analysis for the behavioral sciences. New York: Academic Press 1977.
46. Felder-Puig R, Formann AK, Mildner A et al. Quality of life and psychosocial adjustment of young patients after treatment of bone cancer. *Cancer* 1998, 83, 69-75.
47. Dolgin MJ, Somer E, Buchvald E, Zaizov R. Quality of life in adult survivors of childhood cancer. *Soc Work Health Care* 1999, 28, 31-43.
48. Elkin TD, Phipps S, Mulhern RK, Fairclough D. Psychological functioning of adolescent and young adult survivors of pediatric malignancy. *Med Pediatr Oncol* 1997, 29, 582-588.
49. Teta MJ, Del Po MC, Kasl SV, Meigs JW, Myers MH, Mulvihill JJ. Psychosocial consequences of childhood and adolescent cancer survival. *J Chron Dis* 1986, 39, 751-759.
50. Mackie E, Hill J, Kondryn H, McNally R. Adult psychosocial outcomes in long-term survivors of acute lymphoblastic leukaemia and Wilms' tumour: a controlled study. *Lancet* 2000, 355, 1310-1314.
51. Zevon MA, Neubauer NA, Green DM. Adjustment and vocational satisfaction of patients treated during childhood or adolescence for acute lymphoblastic leukemia. *Am J Pediatr Hematol Oncol* 1990, 12, 454-461.

52. Koocher GP, O'Malley JE, Gogan JL, Foster DJ. Psychological adjustment among pediatric cancer survivors. *J Child Psychol Psychiatry* 1980, 21, 163-173.
53. Chang P-N, Nesbit ME, Youngren N, Robison LL. Personality characteristics and psychosocial adjustment of long-term survivors of childhood cancer. *J Psychosoc Oncol* 1987, 5, 43-58.
54. Dongen-Melman van J. On surviving childhood cancer: Late psychosocial consequences for patients, parents and siblings. Thesis. Erasmus University Rotterdam, The Netherlands 1995.
55. Sloper T, Larcombe IJ, Charlton A. Psychosocial adjustment of five-year survivors of childhood cancer. *J Cancer Educ* 1994, 9, 163-169.
56. Breetveld IS, van Dam FSAM. Underreporting by cancer patients: the case of response-shift. *Soc Sci Med* 1991, 32, 981-987.
57. de Haes JCJM, van Knippenberg FCE. Quality of life of cancer patients: review of the literature. In: Aaronson NK, Beckmann J, eds. The quality of life of cancer patients. New York: Raven Press 1987, 167-182.
58. Aaronson NK. Quality of life research in cancer clinical trials: A need for common rules and language. *Oncology* 1990, 4, 59-66.
59. Padilla GV, Grant M, Ferrell BR. Nursing research into quality of life. *Quality of Life* 1992, 1, 341-348.
60. Patterson JM. Promoting resilience in families experiencing stress. *Ped Clin North America* 1995, 42, 47-63.
61. Sprangers MAG. Response-shift bias: a challenge to the assessment of patients' quality of life in cancer clinical trials. *Cancer Treat Rev* 1996, 22, 55-62.
62. Schwartz CE, Feinberg RG, Jilinskaja E, Applegate JC. An evaluation of a psychosocial intervention for survivors of childhood cancer: paradoxical effects of response shift over time. *Psychooncology* 1999, 8, 344-354.
63. Sprangers MA, van Dam FSAM, Broersen J et al. Revealing response shift in longitudinal research on fatigue: the use of the Thentest approach. *Acta Oncol* 1999, 38, 709-718.
64. Gijsbers van Wijk CMT, Kolk AM. Sex differences in physical symptoms: the contribution of symptom perception theory. *Soc Sci Med* 1997, 45, 231-246.
65. Phillips DL, Segal BE. Sexual status and psychiatric symptoms. *American Sociological Review* 1969, 34, 58-71.
66. Evans DR, Pellizzari JR, Culbert BJ, Metzen ME. Personality, marital, and occupational factors associated with quality of life. *J Clin Psychol* 1993, 49, 477-485.
67. Cope DG. Self-esteem and the practice of breast self-examination. *Western J Nurs Research* 1992, 14, 618-631.
68. Dirksen SR. Perceived well-being in malignant melanoma survivors. *Oncol Nurs Forum* 1989, 16, 353-358.
69. Gotay CC, Muraoka MY. Quality of life in long-term survivors of adult-onset cancers. *J Natl Cancer Inst* 1998, 90, 656-667.
70. Wallace WHB, Blacklay A, Eiser C, Davies H, Hawkins M, Levitt GA, Jenney MEM. Developing strategies for long term follow up of survivors of childhood cancer. *BMJ* 2001, 323, 271-274.

Chapter 7

Educational achievement, employment and living situation in long-term young adult survivors of childhood cancer

Abstract

Purpose: We investigated educational achievement, employment status, living situation, marital status and offspring in 500 Dutch long-term young adult survivors of childhood cancer. The results were compared with a reference group of 1092 persons with no history of cancer. The impact of demographic and medical characteristics on psychosocial adjustment was studied.

Patients and methods: The median age at follow-up of the survivors was 22 years (age range, 16-49 years, 47% female), and that of the comparison group 26 years (age range, 15-53 years, 55% female). All participants completed a self-report questionnaire.

Results: The results showed that, although many survivors are functioning well and leading normal lives, a subgroup of survivors were less likely to complete high-school, to attain an advanced graduate degree, to follow normal elementary or secondary school and had to be enrolled more often on learning disabled programs. The percentage of employed survivors was lower than the percentage of employed controls in the comparison group, but more survivors were student or homemaker. Survivors had lower rates of marriage and parenthood, and worried more about their fertility and the risk of their children having cancer. Survivors, especially males, lived more often with their parents or were single. Cranial irradiation dose ≤ 25 Gy was an important independent prognostic factor of lower educational achievement. Survivors with a history of brain/CNS tumours had a higher risk of being single than survivors with a diagnosis of leukaemia/non-Hodgkin's lymphoma without cranial irradiation.

Conclusion: These results indicate that important psychosocial aspects of life are affected in a substantial number of persons who have been diagnosed with cancer during childhood or adolescence.

Introduction

Treatment advances have led to dramatically improved survival rates for most cancers in children and adolescents. Whereas the proportion of childhood cancer patients who survived 5 or more years was only one in four in 1960, this proportion increased to approximately three in four in 1993 [1]. As the likelihood and period of survival increase, the long-term physical and psychosocial consequences of childhood cancer and its treatment become increasingly important to address. Numerous studies of long-term physical effects have documented the negative impact of surgery, radiation therapy, and chemotherapy involving various organ systems, fertility and reproductive systems, neuropsychological problems, and visible physical impairments [2-5]. Much less research has been undertaken to determine the psychosocial impact of childhood cancer on long-term survivors. Moreover, previous studies into this area have varied in their conclusions. Some of the earliest work suggest that survivors were at increased risk of maladaptive psychosocial sequelae [6]. Other more recent studies report also that survivors show clinical evidence of a moderate degree of emotional difficulty [7-10]. Certain combinations of demographic variables, disease and treatment factors have been suggested as predictors of ultimate psychological adjustment in survivors of childhood cancer. Sex, age at diagnosis, time since completion of therapy and cranial irradiation [11-16] have been among the most consistently cited risk factors. In addition, a number of studies highlighted several difficulties that some survivors experience in adult life, including problems with schooling and employment, insurability, the ability to marry and have children, and in establishing stable relationships with family and friends [10, 17-22]. In contrast, other authors report more positive findings, concluding that most long-term survivors function well psychologically and do not have significant problems [23-26]. It is unclear how to reconcile the findings with the studies done so far. Differences in methodology among previous studies may account for some of the discrepancies. Small numbers of patients and lack of appropriate control groups, for example, limit the conclusions that might be drawn from the majority of these studies [27,28]. However, based on these results, combined with clinical observation and reports of former patients, there is enough reason to assume that at least a subgroup of childhood cancer survivors experience difficulties with overall psychological functioning.

The large majority of psychosocial studies of childhood cancer survivors has been conducted in the United States. Per definition one cannot assume that the results of these studies can be generalised to the European setting. Cultural differences may exist both in public attitudes toward cancer, and in the nature and organisation of the health care and insurance systems. Considering the limitations of the studies done so far and hypothesising that, due to cultural differences, the prevalence and severity of long-term psychosocial problems among childhood cancer survivors in the Netherlands might be different from those in the United States, a study was conducted. The aim of the study was to determine the psychosocial adjustment in a large cohort of Dutch long-term adult survivors of childhood cancer, the so-called Psychosocial Adjustment and Quality Of Life Study in long-term survivors of childhood cancer (PAQOLS). A substantial comparison group of young adults with no history of cancer was recruited to overcome some of the limitations of earlier studies. The findings reported in this paper are part of the PAQOLS mentioned above and we focused on the following questions: 1) What is the level of psychosocial adjustment with respect to educational achievement, employment status, living situation, marital status and offspring in comparison with a non-cancer group and 2) what is the influence of sex, age at follow-up, age at diagnosis, time since completion of therapy, duration of treatment, diagnosis and treatment with or without cranial irradiation on psychosocial adjustment in survivors of childhood cancer?

Patients and methods

Study group
Data were collected from two samples: young adult survivors of childhood cancer (hereafter referred to as the survivors group) and a reference group of persons with no history of cancer (hereafter referred to as the comparison group).

Survivors group
Patients eligible for participation in the study were those attending the long-term follow-up clinic at The Emma Kinderziekenhuis/Academic Medical Center in Amsterdam for their annual evaluation between February 1996 and July 1999. The long-term follow-up clinic was established in 1996 to monitor long-term sequelae of childhood cancer and its treatment. Patients become eligible for transfer from active-treatment clinics to the follow-up clinic when they completed cancer treatment successfully at least 5 years earlier. Survivors are evaluated annually in the clinic by a paediatric oncologist (persons aged <18 years) or internist-oncologist (persons aged >18 years) for late medical effects, as well as a research nurse or psychologist for psychosocial effects. Study participants had to be aged 16 years or older, should have had a pathologic confirmation of malignancy and their cancer had to be diagnosed before the patients were 19 years of age. Within the study period a total of 543 patients were offered appointments to attend the follow-up clinic. Of these patients, 11 attended but were not included in this study (one person was schizophrenic, 8 were developmentally delayed and 2 were deemed ineligible because of a current health problem causing emotional upset). A further 32 patients did not attend, representing a failed attendance rate over this period of 6%. Survivors who did attend were significantly younger than survivors who did not attend (mean = 24 versus 26 years; P < 0.05), but there were no differences with regard to sex or type of diagnosis. A total of 500 patients were approached to take part in this study during their visit at the follow-up clinic and all agreed to participate. After their informed consent the survivors were individually asked to complete a questionnaire by one of the authors (NL). The investigator was present to make sure that the questionnaire was clearly understood. Most survivors had no difficulty in completing the questionnaire, only 3 persons needed some assistance.

Comparison group
Comparison participants were recruited with the help of survivors' general practitioners (GPs). Letters with response cards were send to the GPs of the potential survivor group (n=540) explaining the purpose of the study and asking for their help in selecting an age-matched control group.
Three hundred and thirty GPs responded (61% response rate), of which 151 stated that they could not participate because of a lack of time: thus, 179 from the notified GPs agreed to take part in the study. These GPs were asked to select 10 persons from their patient registry lists (starting with a given letter from the alphabet) with a given sex and age range. Those with (a prior history of) cancer were excluded. The GPs had to sent eligible persons a packet containing the questionnaire, a stamped return address envelope and a cover letter. Two weeks

after the original mailing date the GPs had to sent another packet with the same content and a reminder letter. Of 1790 questionnaires mailed, 23 were returned because the persons had moved and their address could not be traced. Twenty four persons let us know that they refused to participate for various reasons (lack of time n=5; not able to understand Dutch n=4; perceived invasion of privacy n=7; other reasons n=8). Of the 1743 remaining questionnaires 1096 completed questionnaires were returned (response rate 63%). Four responses were excluded afterwards because the respondents did not met the inclusion criteria (too young n=3; too old n=1); thus, 1092 responses were included in the final analysis.

Measures
As no standardised psychosocial instrument was available, we designed a structured self-report questionnaire for the overall study after an extensive literature review related to long-term survivors of childhood cancer and psychosocial functioning. To enhance the face validity of the data, the questions were checked by medical and nursing staff experienced in the care of childhood cancer or in quality of life research. Additionally we piloted the questionnaire on 10 survivors before the study. The final questionnaire covered the following topics: demographic and medical information; educational achievement; occupational status; work and insurance discrimination; quality of life comprising physical, psychological, sexual, and social functioning; experienced health status and worries about health; life style and philosophy of life; relationship with family and friends; self-esteem; fatigue; depression and post-traumatic stress. The questionnaires were identical for both the survivor and the comparison group except for a few questions that specifically addressed the issue of former illness and treatment. In these cases the term 'cancer' was replaced by 'any health problem'. Each questionnaire took 30 to 45 minutes to complete.
For the purposes of the findings reported in this paper, the following items of three sections (with a total of 10 items) were used:

educational achievement
Persons were asked to report 1) their highest level of completed schooling (low education level = less than high-school, high education level = high-school or advanced graduate degrees); 2) enrolment in learning disabled programs.

Employment status
Respondents were asked about 1) their current employment status (employed, unemployed, student/ homemaker, disabled); 2) discrimination because of health history (in job/military service).

living situation, marital status and offspring
Persons were also asked to report 1) their living situation (with parents versus other); 2) marital status (single versus living together/married); 3) having children; 4) health status of children; 5) worries about infertility; 6) worries about health of their children.
In addition, a paediatrician, who had experience regarding the long-term effects of childhood cancer, reviewed the medical record prior to the visit to obtain infor-

mation about survivors' age at diagnosis, time since completion of therapy, duration of treatment, diagnosis, type of treatment, and possible late effects. These
data were recorded onto structured data coding sheets.

Statistical analysis
Differences between the demographic characteristics of the survivor and the
comparison group was analysed with the Chi-square test (categorical variables)
and the unpaired *t*-test (continuous variables). Differences between the psychosocial variables of both groups were adjusted for gender and analysed using the
Mantel-Haenszel summary Chi-square test. No adjustment was made for age. We
dichotomised age into 15-29 years (adolescents and young adults) and age $\geq$ 30
years (adults) and performed a multiple logistic regression analysis with subject
status (comparison group = 0, survivors = 1) as the dependent variable. After an
adjustment for the effects of age and sex and their interaction, age no longer predicted subject status. The univariate associations between the survivors' demographic and medical characteristics on the one hand and the psychosocial variables on the other was calculated using the ordinary Chi-square test. In view of
the explorative characteristic of the study no adjustment for multiple comparison
were made [29]. Finally, all demographic and medical characteristics were presented to logistic regression models to assess their independent impact on psychosocial functioning (lower educational level, living with parents, single). The
independent explanatory values of the characteristics were expressed in adjusted
odds ratios (OR), with their 95% confidence intervals (CI). The corresponding *p*-
values were calculated with the Wald statistic. Calibration of the regression models was assessed with the Hosmer-Lemeshow goodness-of-Fit test [30]. All
analyses were performed with SPSS for Windows, version 9.01.

Results

Demographic and medical characteristics
Information about the demographic and medical characteristics of the survivor
and the comparison group is listed in Table 1. The survivors' age at diagnosis
ranged from 0 to 19 years (median 8), the range of the time since completion of
therapy was 5 to 33 years (median 15). The median duration of treatment was 10
months (range 0-170 months). The survivors were treated for a variety of cancers.
The most frequent diagnoses were leukaemia, non-Hodgkin's lymphoma, Wilms'
tumour and Hodgkin's disease. We compared the distribution of cancer diagnoses
of our study population with the distribution of diagnoses of the survivors, who
were known to be alive but were not yet seen in the follow-up clinic. We found
an over-representation of leukaemias and lymphomas in our study group. For
logistic considerations many survivors with these diagnoses were seen in the follow-up clinic during the first years after the clinic was set up.
The median age of the comparison group was 26 years (range 15-53 years), this
was slightly higher than that of the survivors (median 22; range 16-49 years).
However, when dichotomising age into 15-29 years (adolescents and young
adults) and age $\geq$ 30 years (adults) no difference in frequencies was observed (P
0.90). Further, the differences in proportions of men and women in the groups

were significant: there were more men than women in the survivor group (53% versus 47%), whereas the reverse was true for the comparison group (45% versus 55%).

Educational achievement

In the questionnaire the highest level of education thus far achieved had to be reported. Although some respondents were too young to have reached their final educational level, it was apparent that the educational achievement levels of the survivors were significantly different from those of the comparison group (Table 2). Especially female survivors were less likely to complete high-school or to attain an advanced graduate degree. Significantly more survivors than their controls were unable to follow normal elementary or secondary school and had to be enrolled in learning disabled programs.

Employment status

Table 2 gives also the detailed employment characteristics of both groups. The percentage of employed survivors was significantly lower compared with the comparison group. However, more survivors indicated that they were students or homemakers. Among the employed groups, male survivors were less likely to be employed full-time (85% versus 92%, $P < 0.02$). Especially in male survivors more disability was reported than in the comparison group. Although the numbers were small, the disabilities reported by the survivors included visual handicaps, problems with mobility (amputation/endoprosthesis, scoliosis, hemiplegia), neuropsychological problems (learning difficulties, epilepsy, psychiatric therapy), and extreme fatigue. Survivors also experienced some form of job discrimination as a result of their health history. More than half of the male survivors were rejected for military service.

Table 1. Demographic and medical characteristics of the study group

Variable	Survivors (n=500)		Comparison group (n= 1092)	P-value
Age at diagnosis (years)				
Mean ± SD	8 ± 4.7			
Time since completion of therapy (years)				
Mean ± SD	15 ± 5.8			
Duration of treatment (months)				
Mean ± SD	16 ± 21.4			
Diagnosis	n	%		
Leukaemia	107	21		
Non-Hodgkin's lymphoma	72	14		
Wilms' tumour	64	13		
Hodgkin's disease	62	12		
Soft tissue sarcoma	43	9		
Brain/CNS tumours	45	9		
Osteosarcoma	29	6		
Ewing sarcoma	20	4		
Neuroblastoma	19	4		
Other diagnosis	39	8		
Treatment %				
Chemotherapy, radiation		28		
Chemotherapy, radiation, surgery		19		
Chemotherapy, surgery		22		
Radiation, surgery		4		
Single-modality therapy		27		
Age at follow-up (years)				
Mean ± SD	24 ± 5.1		26 ± 5.2	< 0.001[*]
Sex %				
Men	53		45	0.004[#]
Women	47		55	

[*] t-test
[#] chi-square

Table 2. Educational achievement and employment characteristics

Characteristic	Men		Women		P-value[*]
	survivors (n=267) % (n)	comparison group (n=496)[a] % (n)	survivors (n=233) % (n)	comparison group (n=592)[a] % (n)	
Educational achievement					
Lower educational level	61 (163)	58 (286)	70 (163)	57 (335)	0.003
Higher educational level	39 (104)	42 (210)	30 (70)	43 (257)	0.003
Enrolment in learning disabled programs	9 (25)	3 (15)	6 (15)	2 (13)	< 0.001
Employment characteristics					
Employed	53 (142)	78 (391)	53 (123)	71 (421)	< 0.001
Full-time	85 (121)	92 (360)	65 (80)	65 (274)	0.23
Part-time	15 (21)	8 (31)	35 (43)	35 (147)	0.23
Unemployed	6 (16)	2 (8)	3 (7)	2 (13)	0.005
Student/homemaker	35 (91)	19 (92)	36 (84)	25 (148)	< 0.001
Working in sheltered Workshop	2 (6)	<1 (2)	3 (7)	0 (0)	< 0.001
Disability (50%-100%)	4 (12)	<1 (3)	5 (12)	2 (10)	< 0.001
Discrimination					
Ever denied a job because of cancer history (health problem)?	6 (15)	3 (17)	6 (13)	1 (8)	0.002
Denied entry in military service?	55 (146)	27 (132)	..	..	<0.001[#]

[a] Sample sizes may not always add up to 1092 due to missing values
[*] Mantel-Haenszel summary chi-square
[#] chi-square

Living situation, marital status and offspring
Some significant differences were found between the survivor group and the comparison group regarding living situation, marital status and having children (Table 3). Survivors, especially male survivors, lived more often with their parents and were less frequently married or lived together than controls. The percentage of survivors with biologic children was significantly lower than in the comparison group. The survivor group also reported worrying more about their fertility and the risk of their children having cancer.

Table 3. Living situation, marital status and offspring

Characteristic	Men		Women		P-value[*]
	survivors (n=267) % (n)	comparison group (n=496)[a] % (n)	survivors (n=233) % (n)	comparison group (n=592)[a] % (n)	
Living situation					
With parents	58 (156)	37 (182)	46 (107)	23 (136)	< 0.001
Other	42 (111)	63 (314)	54 (126)	77 (456)	< 0.001
Marital status					
Single	81 (216)	54 (267)	63 (146)	39 (233)	< 0.001
Living together/married	19 (51)	46 (229)	37 (87)	61 (359)	< 0.001
Having biologic children	3 (9)	15 (76)	15 (34)	23 (134)	< 0.001
Children with health problems	0 (0)	5 (4)	6 (2)	5 (7)	0.84
Worries about whether they can have children	53 (142)	29 (144)	65 (151)	55 (324)	< 0.001
Worries about children getting cancer	40 (106)	30 (150)	51 (119)	34 (203)	< 0.001

[a] Sample sizes may not always add up to 1092 due to missing values
[*] Mantel-Haenszel summary chi-square

Survivor characteristics and psychosocial functioning

Table 4 shows the univariate relations between the demographic and medical characteristics of the survivors at the one hand and survivors' educational achievement, living situation and marital status at the other. Because the number of unemployed survivors was very small, we did not analyse this characteristic. Lower educational level was significantly associated with female gender, type of diagnosis and dose of cranial radiation therapy (CRT). Living alone was correlated with all demographic and medical characteristics, with exception of duration of treatment and cranial irradiation. Finally, gender, age at follow-up, age at diagnosis, time since completion of therapy, and diagnosis were significantly associated with marital status.

When the demographic and medical characteristics were entered into the multivariate logistic regression model, the results showed that female survivors were more likely to have a lower educational level than male survivors (Table 5). In contrast, male survivors were more likely to live with their parents and to be single. Survivors who were older at time of follow-up, who were older than 6 years at the time of diagnosis, and survivors where time since completion of therapy passed by more than 15 years were less likely to live with their parents or to be single compared to survivors with opposite characteristics. Survivors with a history of brain/CNS tumours tended to be substantially more likely to be single compared to survivors with a diagnosis of leukaemia/non-Hodgkin's lymphoma without CRT. Finally, cranial irradiation was the strongest independent prognostic factor of educational achievement. Survivors who received a radiation dose of ≤ 25 Gy were about 8 times more likely to have a lower educational level than survivors without this type of treatment.

Discussion

To our knowledge, the current study is the first to evaluate educational achievement, employment status, living situation, marital status and offspring in a large sample of young adult survivors of childhood cancer in The Netherlands compared with a control group. The result show that, although many survivors are functioning well and leading normal lives, some important aspects of life, such as education, living situation and marital status are adversely affected in a subgroup of survivors of childhood cancer. This finding confirms other studies.

Some limitations need to be mentioned. First, as with all survey studies, the likelihood of respondents of the comparison group being those who are particularly interested in the topic is high. For this reason we have to consider that a selection bias in the comparison group was operating in this study. For example, respondents may be those with a higher educational level who may have a stronger social commitment. Second, it should be mentioned that patients attending a long-term follow-up clinic may not be fully representative of the larger population of survivors of childhood cancer. It is possible that survivors participating in the current study might have more somatic or psychological problems than non-compliant survivors, thereby biasing research findings. Third, the survivors in our study were diagnosed over a long period of time, 1963-1992, and almost half of the survivors were treated before 1978. Therefore, they were treated in a variety

Table 4. Association between survivors' characteristics and educational level, living situation and marital status: univariate analysis

Characteristic	Educational level % low (n=326)	% high (n=174)	P-value*	Living situation % with parents (n=263)	% other (n=237)	P-value*	Marital status % single (n=362)	% married/ living together (n=138)	P-value*
Sex									
Men	61	39	0.04	58	42	0.005	81	19	<0.001
Women	70	30		46	54		63	37	
Age at follow-up (yr.)[1]									
<30	64	36	0.32	60	40	<0.001	79	21	<0.001
≥30	29	71		6	94		32	68	
Age at diagnosis (yr.)[1]									
0-6	67	33	0.24	63	37	<0.001	80	20	<0.001
7-12	67	33		50	50		72	28	
≥13	58	42		35	65		57	43	
Time since completion of therapy (yr.)[1]									
≤15	63	37	0.34	65	35	<0.001	82	18	<0.001
≥16	67	33		39	61		61	39	
Duration of treatment (months)[1]									
≤12	63	37	0.28	53	47	0.97	72	28	0.65
≥13	68	32		52	48		73	27	
Diagnosis									
Leukaemia/NHL without CRT (n=135)	59	41	0.001	56	44	0.03	74	26	0.008
Leukaemia/NHL with CRT (n=106)	78	22		62	38		78	22	
Solid tumours (n=214)	59	41		45	55		65	35	
Brain/CNS tumours (n=45)	80	20		53	47		87	13	
Cranial irradiation[1]									
No CRT (n=335)	59	41	<0.001	51	49	0.27	70	30	0.12
≤25 Gy (n=95)	82	18		60	40		77	23	
≥26 Gy (n=70)	74	26		51	49		80	20	

*chi-square

Abbreviations: yr: years, NHL: non-Hodgkin's lymphoma, CRT: cranial radiation therapy

[1]Cut-off point arbitrarily defined on base of clinical experience

Table 5. Impact of survivors' characteristics on educational level, living situation and marital status: multivariate logistic regression model: Adjusted Odds ratio's (OR) and their 95% confidence interval (% CI)

Characteristic	Lower educational level			Living with parents			Single		
	OR	(95% CI)	P-value*#	OR	(95% CI)	P-value*#	OR	(95% CI)	P-value*#
Sex									
Men (r)									
Women	1.47	(0.99 – 2.18)	0.06	0.57	(0.37 – 0.88)	0.01	0.32	(0.19 – 0.51)	<0.001
Age at follow-up (yr.)¹									
<30 (r)									
≥30	1.42	(0.70 – 2.84)	0.32	0.13	(0.04 – 0.38)	<0.001	0.31	(0.15 – 0.66)	0.002
Age at diagnosis (yr.)¹									
0-6 (r)									
7-12	0.96	(0.58 – 1.58)	0.87	0.26	(0.14 – 0.46)	<0.001	0.28	(0.15 – 0.53)	<0.001
≥13	0.65	(0.35 – 1.20)	0.17	0.10	(0.05 – 0.22)	<0.001	0.11	(0.05 – 0.26)	<0.001
Time since completion of therapy (yr.)¹									
≤15 (r)									
≥16	0.93	(0.55 – 1.55)	0.77	0.21	(0.12 – 0.38)	<0.001	0.19	(0.10 – 0.39)	<0.001
Duration of treatment (months)¹									
≤12 (r)									
≥13	0.75	(0.46 – 1.20)	0.23	0.65	(0.38 – 1.10)	0.11	0.99	(0.56 – 1.76)	0.98
Diagnosis									
Leukaemia/NHL without CRT(r)									
Leukaemia/NHL with CRT	0.34	(0.07 – 1.79)	0.21	2.97	(0.48 –18.29)	0.24	2.39	(0.29 –19.77)	0.42
Solid tumours	0.79	(0.48 – 1.31)	0.37	0.66	(0.38 – 1.16)	0.15	0.89	(0.48 – 1.65)	0.07
Brain/CNS tumours	0.88	(0.21 – 3.77)	0.87	1.62	(0.34 – 7.75)	0.54	4.54	(0.88 –23.49)	0.07
Cranial irradiation¹									
No CRT(r)									
≤25 Gy	8.79	(1.63 –47.12)	0.01	0.55	(0.09 – 3.36)	0.52	0.72	(0.09 – 5.80)	0.75
≥26 Gy	2.61	(0.77 – 8.88)	0.12	0.75	(0.19 – 2.87)	0.67	1.22	(0.34 – 4.39)	0.76

* Calculated with Wald Statistics
# Hosmer-Lemeshow goodness-of-fit test (lower educational level p 0.22; living with parents p 0.09; single p 0.92)
Abbreviations: r: reference group; yr: years, NHL: non-Hodgkin's lymphoma, CRT: cranial radiation therapy
¹ Cut-off point arbitrarily defined on base of clinical experience

of ways. When compared with survivors being treated after 1978, survivors treated before this time period received more cranial irradiation and less multiple-agent chemotherapy. Finally, another limitation which should be taken into account are the demographic and medical characteristics included in the logistic regression model. Possible other, not included characteristics (e.g. the perception of impact of diagnosis and treatment), may also have influenced lower educational level, living situation and being single.

Educational achievement. In this study we evaluated only the level of education attained and ascertained the type of school attended, we did not obtain grades or test scores. However, our study showed an impairment of educational achievement in childhood cancer survivors. This finding is not surprising. The literature suggest that educational achievement may be affected by childhood and adolescent cancer in many ways. Firstly, the cancer or the treatment for cancer might cause physical or mental impairments that would make it more difficult to learn. It is well documented that cranial irradiation, intrathecal chemotherapy, high-dose systemic methotrexate, and brain surgery might affect school performance [31-35]. Secondly, in addition to the above mentioned treatments, learning potentials can be also impacted by numerous or lengthy hospitalisations and frequent absenteeism from school. Finally, there may be social factors operating such that parents and teachers tend to alter their attitudes toward the child because of the trauma of the illness and the treat of death [36]. Almost half of our survivors were treated for leukaemia, lymphomas or brain/CNS tumours and many of them received the above mentioned damaging treatment. Although we could not demonstrate a direct relation between diagnosis and educational level, survivors with leukaemia/non-Hodgkin's lymphoma with CRT and survivors with brain/CNS tumours had a lower educational level than survivors with other diagnosis. Our results show also that cranial irradiation was an important explanatory factor of lower educational level. A dose of ≤ 25 Gy had the highest risk. A possible explanation might be that 28% of the children who received ≤ 25 Gy were less than 6 years of age at the time of treatment, whereas only 9% in this age group received ≥ 26 Gy. There is agreement in the literature that younger age at the time of treatment is a risk factor for profound cognitive impairment.

Furthermore, before 1978 less facilities were available in The Netherlands for children to continue their education when in hospital and also skilled school liaisons who helped families with the school system to get the best possible education for their child were not present. Therefore, it is possible that the lower educational level among the survivors may be related to declines in academic ability secondary to prolonged absence from school or to a lack of attention to the child's special education needs. Considering the fact that before 1978 childhood cancer was often a fatal disease, it is likely that decreased academic expectations on the part of parents and teachers might play a role as well. Parents and teachers may have felt that it was enough that the child was well and academic success was of less importance.

Most of the survivors who were enrolled in learning disabled programs were treated for leukaemia, lymphoma or brain/CNS tumours and almost half of them had received treatment with cranial irradiation. As it is well known that children with these diagnoses and treatment are at the greatest risk for learning disabilities it is surprising that the number of survivors who were enrolled in learning disabled programs, although significantly higher compared with the comparison group, was so small. Findings from previous research have suggested sex differences among long-term Acute Lymphoblastic Leukaemia (ALL) survivors treated prophylactically with both radiation and chemotherapy [37-39]. Generally, the literature has suggested that females have a greater risk than males for intellect deficits following treatment for ALL. The results of our study support this finding.

Employment status. Significantly more survivors were unemployed but this is possibly due to the fact that more survivors were students or homemaker. In our questionnaire we did not distinguish students from homemaker so we could not compose separate groups for analysis.

We also found that survivors experienced more job discrimination than the comparison group, also entry into military service was clearly more difficult for survivors than for their male peers. These findings are in line with reports from other studies [15,40-42].

Living situation, marital status and offspring. Survivors and controls differed in terms of living situation and marital status. Both male and female survivors tended to be less often married or lived together compared with their peers. This finding is in agreement with data reported by other authors, where a significantly lower percentage of marriage compared to general population norms was found among survivors [17,21,24,26,41,43,44]. Survivors with a history of brain/CNS tumours had a more than 4 times higher risk to be single compared with survivors with other diagnosis. This is in line with the findings of Byrne and colleagues [17] and Rauck and colleagues [21] who found that survivors of brain/CNS tumours accounted for most of the marriage deficit. To our surprise we found that many survivors, and especially male survivors, still lived with their parents. Although there is a tendency for young adults in the Netherlands to stay longer with their parents and live on one's own at an older age, there is a serious discrepancy between the survivor and the comparison group. Why would survivors of childhood cancer leave home at an older age than their peers? Is it a matter of an ongoing manifestation of prolonged dependency behaviour, initiated by earlier unavoidable dependency on parents during times of illness? From clinical practice, it seems that, due to the former illness, an unusually strong bond tends to develop between the parents and child, sometimes to the extent that they depend exclusively on one another for the satisfaction of emotional and physical needs. This bond may jeopardise the normal process of development of autonomy and the way to independency might be delayed. In this view it is interesting to note that survivors who were very young at time at diagnosis (0-6 years) were also far more likely to live with their parents and to be single in their early adulthood. More extensive exploration of this issue is needed with particular emphasis on the survivor's ability to successfully separate from family.

A lower percentage of survivors compared with controls did not have children. An explanation might be that survivors of childhood cancer were exposed to treatments such as radiation and alkylating agents, which are related to an increased risk of infertility [45]. However, the numbers in this study were too small to identify specific risk factors for infertility. Other reasons for not having children, such as choosing not to have them, were not explored in this study.

Given that a side effect of some cancer treatment is infertility, it is not surprising that the majority of the female survivors reported worrying significantly more about their fertility than their peers. Worries about reproductive capacity were reported in other studies as well [15,23,46]. These studies found also that male survivors reported less often concerns about fertility too. In our study, however, we found that the majority of our male survivors worried about fertility. This is not surprising because cancer treatment can have a negative impact on the fertility of males and females alike.

Although the results of studies looking at the rate of birth defects in children born to childhood cancer survivors are very encouraging [47], survivors worry often about the health of their future children [46,48]. They are afraid that a child conceived after surgery, radiation, or chemotherapy might be born defective or disabled. They also sometimes wonder if they could pass on their cancer genetically to their children. In our study worries about the risk of their children having cancer were also stronger among survivors than peers.

The fact that duration of treatment *per se* was not associated with a lower educational level, living with parents and being single makes sense intuitively. It is more likely that intensity of treatment together with duration of treatment is of importance.

Survivors where time since completion of therapy passed by more than 15 years and older survivors were less likely to live with their parents or to be single. It is not difficult to imagine that the greater the number of years since completion of therapy, the older the survivor will become and the less likely one is to be "affected" by the disease and/or treatment. It is important to note, however, that we have not evaluated person's satisfaction with their living situation, including the quality of social relationships.

In conclusion, the results of this study indicate that a substantial number of persons who have been diagnosed with cancer during childhood or adolescence continue to experience residual limitations in their psychosocial functioning many years after their initial diagnosis and successful medical treatment. The magnitude, the type of these problems and the influencing factors, as measured in The Netherlands, appears to be the same as that in the United States.

Unfortunately, it is not always possible to prevent late effects entirely. The management of a child with cancer is always a balancing act, weighing the need for cure against the risk of late effects.

Our knowledge of the natural history of late effects is evolving slowly and there is still much that is not understood. As the number of survivors increase, it is becoming clear that there is some variability in physical and psychological outcome among childhood cancer survivors. Many young adult survivors of childhood cancer function as well as the general population, and seem to live a normal life with activities common for their age. However, there will be survivors who may continue to experience problems that can affect various aspects of their daily lives many years after treatment. Given the range of outcome, follow-up of survivors is considered essential [3,5,49]. Not only investigation of risk factors and causes of all possible adverse outcomes of childhood cancer and its treatment is important, early identification of late effects and appropriate intervention may be in the interest of the individual survivor. It is possible that some problems can be prevented and others remediated with the provision of appropriate care. Therefore, counselling and advice should be available to facilitate adaptation to such problems and limitations.

Acknowledgments
We are grateful to the people willing to participate in our study.

References

1. End Results Group, 1960-1973. National Cancer Institute's Surveillance, Epidemiology and End results SEER 1997.
2. Schwartz CL, Hobbie WL, Constine LS. Survivors of Childhood Cancer: Assessment and Management. St Louis, Missouri: Mosby-Year Book, Inc 1994.
3. Hawkins MM, Stevens MC. The long-term survivors. *Br Med Bull* 1996, 52, 898-923.
4. Hobbie W, Ruccione K, Moore IK, Truesdell S. Late effects in long-term survivors. In: Foley GV, Fochtman D, Mooney KH, eds. Nursing Care of the Child with Cancer. Orlando, Florida: W.B. Saunders Company 1993 466-496.
5. Wallace WHB, Blacklay A, Eiser C, Davies H, Hawkins M, Levitt GA, Jenney MEM. Developing strategies for long term follow up of survivors of childhood cancer. *BMJ* 2001, 323, 271-274.
6. Koocher GP, O'Malley JE, Gogan JL, Foster DJ. Psychological adjustment among pediatric cancer survivors. *J Child Psychol Psychiatry* 1980, 21, 163-173.
7. Chang P-N, Nesbit ME, Youngren N, Robison LL. Personality characteristics and psychosocial adjustment of long-term survivors of childhood cancer. *J Psychosoc Oncol* 1987, 5, 43-58.
8. Dongen-Melman van J. On surviving childhood cancer: Late psychosocial consequences for patients, parents and siblings. Thesis. Erasmus University Rotterdam, The Netherlands 1995.
9. Sloper T, Larcombe IJ, Charlton A. Psychosocial adjustment of five-year survivors of childhood cancer. *J Cancer Educ* 1994, 9, 163-169.
10. Zeltzer LK. Cancer in adolescents and young adults psychosocial aspects. Long-term survivors. *Cancer* 1993, 71, 3463-3468.
11. Fobair P, Hoppe RT, Bloom J, Cox R, Varghese A, Spiegel D. Psychosocial problems among survivors of Hodgkin's disease. *J Clin Oncol* 1986, 4, 805-814.
12. Koocher GP, O'Malley JE. The Damocles syndrome: Psychosocial consequences of surviving childhood cancer. New York: McGraw-Hill 1981.
13. Mulhern RK, Ochs J, Fairclough D, Wasserman AL, Davis KS, Williams JM. Intellectual and academic achievement status after CNS relapse: a retrospective analysis of 40 children treated for acute lymphoblastic leukemia. *J Clin Oncol* 1987, 5, 933-940.
14. Rowland JH, Glidewell OJ, Sibley RF, Holland JC, Tull R, Berman A, Brecker ML, Harris M, Glickman AS, Forman E. Effects of different forms of central nervous system prophylaxis on neuropsychologic function in childhood leukemia. *J Clin Oncol* 1984, 2, 1327-1335.
15. Wasserman AL, Thompson EI, Wilimas JA, Fairclough DL. The psychological status of survivors of childhood/adolescent Hodgkin's disease. *AJDC* 1987, 141, 626-631.
16. Williams JM, Davis KS. Central nervous system prophylactic treatment for childhood leukemia: neuropsychological outcome studies. *Cancer Treat Rev* 1986, 13, 113-127.
17. Byrne J, Fears TR, Steinhorn SC, Mulvihill KJJ, Connelly RR, Austin DF, Holmes GF, Holmes FF, Latourette HB, Teta MJ. Marriage and divorce after childhood and adolescent cancer. *JAMA* 1989, 262, 2693-2699.
18. Evans SE, Radford M. Current lifestyle of young adults treated for cancer in childhood. *Arch Dis Child* 1995, 72, 423-426.
19. Haupt R, Fears TR, Robison LL, Mills JL, Nicholson HS, Zeltzer LK, Meadows AT, Byrne J. Educational attainment in long-term survivors of childhood acute lymphoblastic leukemia. *JAMA* 1994, 272, 1427-1432.
20. Kelaghan J, Myers MH, Mulvihill JJ, Byrne J, Connelly RR, Austin DF, Strong LC, Meigs JW, Latourette HB, Holmes GF. Educational achievement of long-term survivors of childhood and adolescent cancer. *Med Pediatr Oncol* 1988, 16, 320-326.
21. Rauck AM, Green DM, Yasui Y, Mertens A, Robison LL. Marriage in the survivors of childhood cancer: a preliminary description from the Childhood Cancer Survivor Study. *Med Pediatr Oncol* 1999, 33, 60-63.
22. Hays DM, Landsverk J, Sallan SE, Hewett KD, Patenaude AF, Schoonover D, Zilber SL, Ruccione K, Siegel SE. Educational, occupational, and insurance status of childhood cancer survivors in their fourth and fifth decades of life. *J Clin Oncol* 1992, 10, 1397-1406.
23. Gray RE, Doan BD, Shermer P, FritzGerald AV, Berry MP, Jenkin D, Doherty MP. Psychologic adaptation of survivors of childhood cancer. *Cancer* 1992, 70, 2713-2721.
24. Green DM, Zevon MA, Hall B. Achievement of life goals by adult survivors of modern treatment

for childhood cancer. *Cancer* 1991, 67, 206-213.

25. Kazak AE. Implications of survival: Pediatric Oncology Patients and their Families. In: Bearison A, Mulhern R, eds. Pediatric Psychooncology. New York: Oxford University Press 1994, 171-192.

26. Meadows AT, McKee L, Kazak AE. Psychosocial status of young adult survivors of childhood cancer: a survey. *Med Pediatr Oncol* 1989, 17, 466-470.

27. Eiser C. Practitioner review: long-term consequences of childhood cancer. *J Child Psychol Psychiatry* 1998, 39, 621-633.

28. Eiser C, Havermans T. Long term social adjustment after treatment for childhood cancer. *Arch Dis Child* 1994, 70, 66-70.

29. Rothman KJ. No adjustments are needed for multiple comparisons. *Epidemiology* 1990, 1, 43-46.

30. Hosmer DW, Lemeshow S. Applied logistic regression. New York: Jon Wiley & Sons 1989.

31. Eiser C. Effects of chronic illness on intellectual development. A comparison of normal children with those treated for childhood leukaemia and solid tumours. *Arch Dis Child* 1980, 55, 766-770.

32. Lansky SB, Cairns NU, Lansky LL, Cairns GF, Stephenson L, Garin G. Central nervous system prophylaxis. Studies showing impairment in verbal skills and academic achievement. *Am J Pediatr Hematol Oncol* 1984, 6, 183-190.

33. Meadows AT, Gordon J, Massari DJ, Littman P, Fergusson J, Moss K. Declines in IQ scores and cognitive dysfunctions in children with acute lymphocytic leukaemia treated with cranial irradiation. *Lancet* 1981, 2, 1015-1018.

34. Mulhern RK, Kovnar E, Langston J, Carter M, Fairclough D, Leigh L, Kun LE. Long-term survivors of leukemia treated in infancy: factors associated with neuropsychologic status. *J Clin Oncol* 1992, 10, 1095-1102.

35. Twaddle V, Britton PG, Craft AC, Noble TC, Kernahan J. Intellectual function after treatment for leukaemia or solid tumours. *Arch Dis Child* 1983, 58, 949-952.

36. Zwartjes WJ. The psychological costs of curing the child with cancer. In: van Eys J, Sullivan MP, eds. Status of the Curability of Childhood Cancers. New York: Raven Press, 1980, 277-284.

37. Brown RT, Madan-Swain A. Cognitive, neuropsychological, and academic sequelae in children with leukemia. *J Learn Disabil* 1993, 26, 74-90.

38. Robison LL, Nesbit ME, Jr., Sather HN, Meadows AT, Ortega JA, Hammond JM. Factors associated with IQ scores in long-term survivors of childhood acute lymphoblastic leukemia. *Am J Pediatr Hematol Oncol* 1984, 6, 115-121.

39. Waber DP, Gioia G, Paccia J, Sherman B, Dinklage D, Sollee N, Urion DK, Tarbell NJ, Sallan SE. Sex differences in cognitive processing in children treated with CNS prophylaxis for acute lymphoblastic leukemia. *J Pediatr Psychol* 1990, 15, 105-122.

40. Hays DM. Adult survivors of childhood cancer. Employment and insurance issues in different age groups. *Cancer* 1993, 71, 3306-3309.

41. Makipernaa A. Long-term quality of life and psychosocial coping after treatment of solid tumours in childhood. A population-based study of 94 patients 11-28 years after their diagnosis. *Acta Paediatr* 1989, 78, 728-735.

42. Tebbi CK, Bromberg C, Piedmonte M. Long-term vocational adjustment of cancer patients diagnosed during adolescence. *Cancer* 1989, 63, 213-218.

43. Teeter MA, Holmes GE, Holmes FF, Baker AB. Decisions about marriage and family among survivors of childhood cancer. *J Psychosoc Oncol* 1987, 5, 59-68.

44. Zeltzer LK, Chen E, Weiss R, Guo MD, Robison LL, Meadows AT, Mills JL, Nicholson HS, Byrne J. Comparison of psychologic outcome in adult survivors of childhood acute lymphoblastic leukemia versus sibling controls: a cooperative Children's Cancer Group and National Institutes of Health study. *J Clin Oncol* 1997, 15, 547-556.

45. Nicholson HS, Byrne J. Fertility and pregnancy after treatment for cancer during childhood or adolescence. *Cancer* 1993, 71, 3392-3399.

46. Weigers ME, Chesler MA, Zebrack BJ, Goldman S. Self-reported worries among long-term survivors of childhood cancer and their peers. *J Psychosoc Oncol* 1998, 16, 1-23.

47. Sankila R, Olsen JH, Anderson H, Garwicz S, Glattre E, Hertz H, Langmark F, Lanning M, Moller T, Tulinius H. Risk of cancer among offspring of childhood-cancer survivors. Association of the Nordic Cancer Registries and the Nordic Society of Paediatric Haematology and Oncolo-

gy. *N Engl J Med* 1998, 338, 1339-1344.

48. Zevon MA, Neubauer NA, Green DM. Adjustment and vocational satisfaction of patients treated during childhood or adolescence for acute lymphoblastic leukemia. *Am J Pediatr Hematol Oncol* 1990, 12, 454-461.

49. Oeffinger KC, Eshelman DA, Tomlinson GE, Buchanan GR, Foster BM. Grading of late effects in young adult survivors of childhood cancer followed in an ambulatory adult setting. *Cancer* 2000, 88, 1687-1695.

Chapter 8

Post Traumatic Stress Symptoms in adult survivors of childhood cancer

Abstract

Purpose: Previous research suggest that posttraumatic stress disorder (PTSD) is present in survivors of childhood cancer. The aim of the current study was to explore posttraumatic stress symptoms in a sample of young adult survivors of childhood cancer. In addition, the impact of demographic, medical and treatment factors on survivors' posttraumatic stress symptoms was studied.

Patients and methods: Participants were 500 long-term survivors of childhood cancer. The median age at follow-up was 24 years (age range, 16-49 years, 47% female). To assess symptoms of posttraumatic stress, all participants completed the Impact of Event Scale (IES), a self-report instrument consisting of two sub-scales, Intrusion and Avoidance.

Results: Twelve percent of this sample of adult survivors of childhood cancer had scores in the severe range, indicating they are unable to cope with the impact of their disease and need professional help. Twenty percent of the female survivors had scores in the severe range as compared with 6% of the male survivors. Linear regression models revealed that being female, unemployed, a lower educational level, type of diagnosis and severe late effects/health problems were associated with posttraumatic stress symptoms.

Conclusion: The results indicate that a substantial subset of childhood cancer survivors report symptoms of posttraumatic stress. This finding support the outcomes reported previously that diagnosis and treatment for childhood cancer may have significant long-term effects, which are manifest in symptoms of posttraumatic stress. The investigated factors could explain posttraumatic stress symptoms only to a limited extent. Further research exploring symptoms of posttraumatic stress in childhood cancer survivors in more detail is clearly warranted. From a clinical perspective, health care providers must pay attention to these symptoms during evaluations in the follow-up clinic. Early identification of PTSD symptoms can enhance the quality of life for survivors of childhood cancer.

Introduction

The topic of long-term consequences of childhood cancer is not new. Many studies have focused on various aspects of psychological well-being, including emotional functioning, depression and anxiety, and self-esteem [1-13]. Besides, social functioning is also studied frequently, covering issues as education, employment, insurance coverage, living situation, marital status, and fertility [1,3,4,6-10,12-30].
Another outcome variable that has been recently studied in childhood cancer is

posttraumatic stress disorder (PTSD). PTSD is considered one of the anxiety disorders. It is characterised by symptoms that can be grouped into three clusters: reexperiencing, arousal and avoidance [31]. To receive a diagnosis of PTSD , one must have been exposed to a traumatic event, defined as imminent threat to life or a serious injury. Furthermore, the person must have manifested some psychological reaction, usually fear, to this event. Subsequently, a constellation of symptoms develops, such as nightmares, intrusive memories of the event, avoidance of trauma-related stimuli, constricted affect, anger, and an exaggerated response. Several authors identified a cluster of anxiety and avoidance symptoms in paediatric cancer survivors and their parents. These symptoms were consistent with a trauma response and have led researchers to propose that the long-term psychosocial impact of cancer may best be understood by using the framework of posttraumatic stress [32-34].

The framework of posttraumatic stress in childhood cancer survivors makes sense, given the potentially traumatic nature of the cancer experience [35]. The threat to life, intensive treatment regimens, painful invasive procedures, and dangerous complications may compound these extremely stressful experiences. In addition, long-term effects of treatment, such as growth retardation, cognitive impairment, physical changes such as amputation, and infertility can serve as life-long reminders [36].

Nir [37] was the first to suggest that children undergoing treatment for cancer might experience posttraumatic stress. Subsequent research have reported symptoms of posttraumatic stress in young children following bone marrow transplant [34,38], in young survivors with paediatric leukaemia [33,39] and in young survivors with a variety of childhood cancers [40,41]. In a study of young adult survivors of childhood cancer, one-fifth of the sample met criteria for a diagnosis of PTSD, with clinically significant symptoms of intrusion and avoidance reported [35].

As with many somatic and psychological long-term consequences of childhood cancer, it is not clear why some children develop posttraumatic stress symptoms after cancer treatment and others do not.

Therefore, it is important that health care professionals are able to identify survivors at greatest risk and specific variables associated with risk for PTSD, and that they also are able to recognise those in greatest need of preventive or rehabilitative services.

The aim of the current study was to explore posttraumatic stress symptoms in a sample of young adult survivors of childhood cancer. In addition, the impact of demographic, medical and treatment factors on survivors' posttraumatic stress symptoms was studied.

Patients and methods

Study group
Patients eligible for participation in the study were those attending the long-term follow-up clinic at The Emma Kinderziekenhuis/Academic Medical Center in Amsterdam for their annual evaluation between February 1996 and July 1999. The long-term follow-up clinic was established in 1996 to monitor long-term

sequelae of childhood cancer and its treatment. Patients become eligible for transfer from active-treatment clinics to the follow-up clinic when they had succesfully completed their cancer treatment at least 5 years earlier. Survivors are evaluated annually in the clinic by a paediatric oncologist (persons aged <18 years) or internist-oncologist (persons aged >18 years) for late medical effects, as well as a research nurse or psychologist for psychosocial effects. Study participants had to be aged 16 years or older, have had a pathologic confirmation of malignancy and their cancer had to have been diagnosed before the patients were 19 years of age. Within the study period a total of 543 patients were offered appointments to attend the follow-up clinic. Of these patients, 11 attended but were not included in this study (one person was schizophrenic, 8 were developmentally delayed and 2 were deemed ineligible because of a current health problem causing emotional upset). A further 32 patients did not attend, representing a failed attendance rate over this period of 6%. Survivors who did attend were significantly younger than survivors who did not attend (mean = 24 versus 26 years; $P < 0.05$), but there were no differences with regard to sex or type of diagnosis. A total of 500 patients were approached to take part in this study during their visit at the follow-up clinic and all agreed to participate. After their informed consent the survivors were individually asked to complete a questionnaire by one of the authors (NL). The investigator was present to make sure that the questionnaire was clearly understood. Most survivors had no difficulty in completing the questionnaire, only 3 persons needed some assistance.

Measures

Data were collected on sociodemographic characteristics in terms of age at follow-up, gender, marital status (single, living together/married), educational level (low = less than high school, high = high-school or advanced degree), and employment status (unemployed, student/homemaker, employed). In addition, a paediatrician, who had experience regarding the long-term effects of childhood cancer, reviewed the medical record prior to the visit to obtain information about survivor's cancer history and these data were recorded onto structured data coding sheets. Age at diagnosis, time since completion of therapy and duration of treatment were assessed. Diagnoses were categorised into: leukaemia/non-Hodgkin's lymphoma with or without cranial radiation therapy (CRT), solid tumours, and brain/central nervous system (CNS) tumours. Treatment was aggregated into three categories: chemotherapy (with or without surgery), radiation therapy (with or without surgery), and combination therapy (chemotherapy and radiation therapy with or without surgery). Finally, late effects and health problems were scored on an adapted version of the Greenberg, Meadows & Kazak's Scale for Medical Limitations [42] by two paediatric oncologists and a paediatric oncology nurse, who were blinded to the survivors' identity. Patients were categorised into the following three groups according to their most serious medical limitation: 1 (mild) = no limitations of activity. This group included children with one kidney and second benign neoplasm's, and no cosmetic or organ dysfunction; 2 (moderate) = no serious restriction of daily life. This group included children with hypoplasia or asymmetry of soft tissue, mild scoliosis and other mild orthopaedic problems, moderate obesity, abnormally short stature, mild hearing loss, cataract, hypothyroidism, delayed sexual maturation, learning delay, enucleation of one eye, small testis, elevated follicle stimulating hormone/luteinising hormone (FSH/LH), alopecia, hypertension, pulmonary diffusion disturbances; 3 (severe) = significant restriction on daily activity or severe cosmetic changes. This group included children with learning delay requiring special education, soft tissue or bone changes that alter appearance, severe asymmetry, absent limb, dental reconstruction, gonadal failure, azoospermia, known sterility, blindness, organ damage that limits activity, second malignant neoplasm's, hemipareses, fatigue that affect daily and social activities. Seventeen percent (n =86) of survivors were categorised as mild, 45% (n = 225) as moderate, and 38% (n =189) as severe.

We used the Impact of Event Scale (IES) to assess symptoms of posttraumatic stress. This questionnaire was developed to assess the level of posttraumatic stress symptoms [43]. It consists of 15 items that are rated on a four-point scale for frequency of occurrence during the previous week and has two subscales, Intrusion and Avoidance. An item in the Intrusion subscale is for instance "I had dreams about it", and an item in the Avoidance scale is among others " I knew that a lot of unresolved feelings were still there, but I kept them under wraps". A score between 8 and 25 means that a person is at risk for developing PTSD and a score of 26 or more is considered to indicate that a person is unable to cope with the impact of the event and needs professional help [44]. The IES has high internal consistency and test-retest reliability; it discriminates between different populations and symptom levels [45], in medical and nonmedical samples [46]. Cronbach's alpha coefficient in this study was 0.88 and 0.84 for the two subscales of intrusion and avoidance, and 0.91 for the total scale, respectively.

Statistical analysis

The Statistical Package for Social Sciences (SPSS) Windows version 9.01 was used for all statistical analyses. Descriptive statistics were performed for all of the variables. Percentual and mean IES scores of the male and female survivors were analysed with Chi-square tests or Student's t-test. To examine the magnitude of these differences, effect sizes were calculated by dividing the difference between a given mean score of the male survivors and the mean score in the female survivor group by the standard deviation of scores in the female survivor group. An effect size of 0.20 is considered small, whereas effect sizes about 0.50 and 0.80 or greater are moderate and large [47].

To investigate which variables predict survivors' posttraumatic stress symptoms, all variables were stepwise presented to a multiple linear regression model to assess their independent prognostic value. For three variables (employment status, diagnosis, and treatment) we created dummy variables and took the first category (unemployment, leukaemia/non-Hodgkin's lymphoma without CRT, and chemotherapy with or without surgery) as reference for the analysis. First (Step 1), survivors' demographic characteristics were presented to the model. Second (Step 2), medical and treatment characteristics were entered. Finally, the demographic, medical and treatment characteristics were entered in the regression together (Step 3). With this strategy, the contribution of the separate steps show which variables contribute especially to posttraumatic stress. For each regression, the explained variance (R square) was determined.

Results

Characteristics of survivor group

Information about the demographic, medical and treatment characteristics of the survivor group is listed in Table 1. The survivors' age at follow-up ranged from 16 to 49 (median 24), the range of age at diagnosis was 0 to 19 years (median 8). Time since completion of therapy ranged from 5 to 33 years (median 15), and the median duration of treatment was 10 months (range 0-170 months). The survivors were treated for a variety of cancers. The most frequent diagnoses were leukaemia, non-Hodgkin's lymphoma, Wilms' tumour and Hodgkin's disease. We compared the distribution of cancer diagnoses of our study population with the distribution of diagnoses of the survivors, who were known to be alive but were not yet seen in the follow-up clinic. We found an over-representation of leukaemias and lymphomas in our study group. From logistic considerations many survivors with these diagnoses were seen in the follow-up clinic during the first years after the clinic was set up.

Posttraumatic stress symptoms in survivors

In Table 2 the mean scores and standard deviations for the IES are presented for the survivors as a group and for male and female survivors. 28% of the sample (n=137) had scores in the moderate range (a total score of 8-25), and 12% (n=62) had posttraumatic stress symptom scores in the severe range (a total score of 26 or more).

Women had significantly higher scores on the scales than men (range effect sizes -0.22 - -0.53). Of the male survivors, 25% (n=65), and 28% (n= 66) of the female survivors had posttraumatic stress symptom scores in the moderate range. Six percent (n=16) of the male survivors and 20% (n=46) of the female survivors had scores in the severe range.

Table 1. Demographic, medical and treatment characteristics of the study group

Variable	Survivors (n=500)	
Age at follow-up (years)		
Mean ± SD	24 ± 5.1	
Sex %		
Men	53	
Women	47	
Marital status %		
Single	72	
Living together/married	27	
Educational status %		
Lower level	65	
Higher level	35	
Employment status %		
Unemployed	10	
Student/homemaker	35	
Employed	55	
Age at diagnosis (years)		
Mean ± SD	8 ± 4.7	
Time since completion of therapy (years)		
Mean ± SD	15 ± 5.8	
Duration of treatment (months)		
Mean ± SD	16 ± 21.4	
Diagnosis	N	%
Leukaemia/non-Hodgkin's lymphoma without CRT	136	27
Leukaemia/non-Hodgkin's lymphoma with CRT	105	21
Solid tumour	214	43
Brain/CNS tumour	45	9
Treatment		
Chemotherapy (with or without surgery)	226	45
Radiation therapy (with or without surgery)	40	8
Combination therapy (chemotherapy and radiation therapy with or without surgery)	234	47
Medical limitations .	N	%
None/mild	86	17
Moderate	225	45
Severe	189	38

Table 2. Mean (SD) scores for the Impact of Event Scale (IES) for the survivors as a group and for the male and female survivors

Scale	Survivors (n=500)	Males (n=266)	Females (n=234)	P-value	effectsize
IES intrusion	5.6 (7.3)	3.4 (5.0)	8.0 (8.7)	< 0.001[*]	-0.53
IES avoidance	4.7 (7.1)	3.9 (6.0)	5.7 (8.1)	0.004[*]	-0.22
IES total symptoms	10.3 (13.3)	7.3 (9.9)	13.8 (15.7)	< 0.001[*]	- 0.41
	n %	n %	n %		
Total score 8-25	137 28	65 25	66 28	0.34	
Total score ≥ 26	62 12	16 6	46 20	< 0.001[#]	

[*] t-test
[#] Chi-square

Prediction of survivors' post-traumatic stress by demographic, medical and treatment characteristics

Table 3 presents the results of predictors of posttraumatic stress symptoms by multivariate linear regression analysis at each step. With regard to the demographic variables (Step 1), female gender was the strongest prognostic factor on posttraumatic stress symptoms. In addition, posttraumatic stress was significantly explained by lower educational level and unemployment.

With regard to the medical and treatment characteristics (Step 2), severe late effects/health problems were associated with the level of posttraumatic stress symptoms.

After entering both demographic, medical and treatment characteristics into the model (Step 3), the results showed that female gender was the strongest predictor of posttraumatic stress symptoms. The level of posttraumatic stress symptoms was further explained by lower educational level, unemployment, survivors who have had leukaemia/non-Hodgkin's lymphoma without CRT, and severe late effects/health problems. The selected characteristics explained only a small proportion of the variability (R^2) of the posttraumatic stress scores, namely 13%.

Discussion

The results and conclusion of the present study should be viewed as preliminary. The current results are hampered by several weaknesses, such as the absence of a structured diagnostic interview with the survivor and the lack of normal control subjects. However, this study addressed a population that has not been the focus of much research and requires clinical attention. Therefore, we felt that the present results are meaningful and contribute to co-operation between clinical work and research.

The findings indicate that a substantial subset of childhood cancer survivors report symptoms of posttraumatic stress (given their cut-off scores on the IES) and support the outcomes reported previously that diagnosis and treatment for childhood cancer may have significant long-term effects, which are manifest in symptoms of posttraumatic stress. Furthermore, the findings are in line with our clinical practice, where a subgroup of survivors report having distressing memo

Table 3. Simultaneous linear regressions (Beta) for survivors' posttraumatic stress. Within each step variables are presented in order of selection, see also Methods section.

	IES total
Step 1	
Demographic characteristics	
Sexe (female)	0.22**
Age at follow-up (years)	0.02
Marital status (married)	0.07
Educational level (higher level)	- 0.14*
Employment status[b]	
Student/homemaker	- 0.21*
Employed	- 0.18*
Total R^{2a}	10%
Step 2	
Medical and treatment characteristics	
Age at diagnosis (years)	0.06
Diagnosis[c]	
Leukaemia/non-Hodgkin's lymphoma with CRT	- 0.09
Solid tumour	- 0.02
Brain/CNS tumour	0.00
Duration of treatment (months)	0.08
Years since completion of therapy	- 0.01
Late effects/health problems	0.16*
Treatment[d]	
Radiation therapy (with or without surgery)	- 0.04
Combination therapy (with or without surgery)	0.02
Total R^{2a}	3%
Step 3	
Demographic, medical and treatment characteristics	
Sexe (female)	0.23**
Age at follow-up (years)	- 0.35
Marital status (married)	0.05
Educational level (higher level)	- 0.15*
Employment status[b]	
Student/homemaker	- 0.20*
Employed	- 0.17*
Age at diagnosis (years)	0.35
Diagnosis[c]	
Leukaemia/non-Hodgkin's lymphoma with CRT	- 0.12*
Solid tumour	- 0.07
Brain/CNS tumour	- 0.07
Duration of treatment (months)	0.06
Years since completion of therapy	0.35
Late effects/health problems	0.15*
Treatment[d]	
Radiation therapy (with or without surgery)	- 0.04
Combination therapy (with or without surgery)	- 0.00
Total R^{2a}	13%

Abbreviations: IES: Impact of Event Scale, CRT: cranial radiation therapy, CNS: central nervous system
[a] R is the percentage of the total variation of the dependent variable score that is explained by the independent variables together
[b] reference group = unemployment, [c] reference group = leukaemia/non-Hodgkin's lymphoma without CRT,
[d] reference group = chemotherapy (with or without surgery)
* Statistically significant differences (p < 0.05)
** Statistically significant differences (p < 0.001)

ries or dreams about aspects of cancer or treatment, feelings that the trauma is recurring (flashbacks), and becoming very upset, scared or angry when talking or thinking about cancer and treatment.

However, when one looks at the incidence of PTSD in the general community and in at-risk groups, it does not appear to occur at an increased incidence in survivors. The incidence of PTSD in the general community is estimated at 1-14%, and the epidemiological rates range from 3-58% in at-risk populations [31,48].

Mean scores of posttraumatic stress symptoms for this sample were also compared with data from a young adult childhood cancer survivors sample from the North-American study by Hobbie and colleagues [35]. The survivors in our study had significantly lower scores on both the Intrusion and Avoidance scale (95% CI of the difference-4.57 to -1.03 and -7.20 to -3.60). The 12% occurrence of high scores in our sample was also significantly lower than the 21% documented in their study. It is not clear why rates of posttraumatic stress in our sample were lower than those seen in their population. Cultural differences may play a role here.

Of all the demographic, medical and treatment factors that were investigated, only females, lower educational level, unemployment, survivors who have had leukaemia/non-Hodgkin's lymphoma without CRT and severe late effects/health problems were associated with posttraumatic stress symptoms. It is difficult to determine whether the present results are reliable because not much research has addressed the associations between these factors in childhood cancer survivors and posttraumatic stress.

Girls and women have been noted to report more symptoms of posttraumatic stress than boys or men with similar exposure [49]. It is known from other studies that women are simply more inclined to actually report the symptoms they experience than men, both in health surveys and in the consultation room [50]. Raised with the "boys don't cry" doctrine, men might be more reserved about their health problems, whereas women might show a greater willingness to discuss symptoms with others [51]. It is perhaps also possible that seemingly similar events are experienced differently by men and women and that as such women have an increased vulnerability to posttraumatic stress.

The fact that severe late effects/health problems were associated with posttraumatic stress makes sense intuitively. It is possible that some of these survivors experience a great deal more anxiety or perceive their life as currently threatened. However, our findings are not consistent with the study by Hobbie and colleagues [35], who did not find an association between medical late effects and scores on the IES.

Unemployment was also found to be associated with posttraumatic stress. It should be noted, however, that the unemployed group was small (n= 48) and more than half of these survivors were, partly or fully, officially declared unfit for work because of medical problems. As survivors with severe late effects/health problems had a higher risk of having posttraumatic stress symptoms, it makes sense that unemployed survivors also report higher levels of stress.

Why survivors with a lower educational level and survivors who have had leukaemia/non-Hodgkin's lymphoma without CRT were more likely to have posttraumatic stress symptoms than survivors with other diagnosis is unclear and should be further explored.

The data presented in this study provides evidence that there is a subgroup of childhood cancer survivors who experience posttraumatic stress symptoms, even years after their experience with cancer and treatment. It is evident that additional research is needed and the importance of understanding symptoms of posttraumatic stress in more detail is clearly warranted. Future research efforts should also be directed at further assessment of posttraumatic stress symptoms using diagnostic psychiatric interviews and for studies examining a wider range of factors, such as preexisiting psychological problems.

The findings should be kept in mind by physicians, nurses, and other health care providers in oncology follow-up clinics. The annual evaluation should include an exploration of symptoms of posttraumatic stress, with a goal of assessing the degree to which these symptoms actually influence adaptive functioning [52]. Although symptoms of PTSD are readily identifiable, the diagnosis is easily missed unless specific inquiries are made. Often practitioners are reluctant to ask their patients about events that are distressing, and most patients will not usually mention such topics without prompting. By providing patients with the opportunity to disclose such events, practitioners break down an important barrier to treatment by legitimising the event as a valid explanation for symptoms [53]. Those individuals exhibiting symptoms of posttraumatic stress should be referred to a psychologist or psychiatrist for a thorough evaluation and intervention. Early recognition can lead to treatments that have found to be effective in treating PTSD, such as exposure therapy (helping patients confront painful memories and feelings), cognitive therapy (helping patients process their thoughts and beliefs), anxiety management, interpersonal therapies (helping patients understand the ways in which the event continues to affect relationships and other aspects of their lives), and group therapy [53]. Early identification of PTSD symptoms can enhance the quality of life for survivors of childhood cancer.

Acknowledgements
We would like to express our general thanks to the persons who contributed to this study. We also thank Prof. dr. B.P.R. Gersons, psychiatrist, for his thoughtful comments.

References

1. Tebbi CK, Bromberg C, Piedmonte M. Long-term vocational adjustment of cancer patients diagnosed during adolescence. *Cancer* 1989, 63, 213-218.
2. Zevon MA, Neubauer NA, Green DM. Adjustment and vocational satisfaction of patients treated during childhood or adolescence for acute lymphoblastic leukemia. *Am J Pediatr Hematol Oncol* 1990, 12, 454-461.
3. Gray RE, Doan BD, Shermer P, FitzGerald AV, Berry MP, Jenkin D, Doherty MA. Psychologic adaptation of survivors of childhood cancer. *Cancer* 1992, 70, 2713-2721.
4. Moe PJ, Holen A, Glomstein A, Madsen B, Hellebostad M, Stokland T, Wefring KW, Steen-Johnson J, Nielsen B, Howlid H, Borsting S, Hapnes C. Long-term survival and quality of life in patients treated with a national all protocol 15-20 years earlier: IDM/HDM and late effects? *Ped Hemat Oncol* 1997, 14, 513-524.
5. Elkin TD, Phipps S, Mulhern RK, Fairclough D. Psychological functioning of adolescent and young adult survivors of pediatric malignancy. *Med Pediatr Oncol* 1997, 29, 582-588.
6. Felder-Puig R, Formann AK, Mildner A, Bretschneider W, Bucher B, Windhager R, Zoubek A, Puig S, Topf R. Quality of life and psychosocial adjustment of young patients after treatment of bone cancer. *Cancer* 1998, 83, 69-75.
7. Dolgin MJ, Somer E, Buchvald E, Zaizov R. Quality of life in adult survivors of childhood cancer. *Soc Work Health Care* 1999, 28, 31-43.
8. Lansky SB, List MA, Ritter-Sterr C. Psychosocial consequences of cure. *Cancer* 1986, 58, 529-533.
9. Zeltzer LK, Chen E, Weiss R, Guo MD, Robison LL, Meadows AT, Mills JL, Nicholson HS, Byrne J. Comparison of psychologic outcome in adult survivors of childhood acute lymphoblastic leukemia versus sibling controls: a cooperative Children's Cancer Group and National Institutes of Health study. *J Clin Oncol* 1997, 15, 547-556.
10. Teta MJ, Del Po MC, Kasl SV, Meigs JW, Myers MH, Mulvihill JJ. Psychosocial consequences of childhood and adolescent cancer survival. *J Chron Dis* 1986, 39, 751-759.
11. Mackie E, Hill J, Kondryn H, McNally R. Adult psychosocial outcomes in long-term survivors of acute lymphoblastic leukaemia and Wilms' tumour: a controlled study. *Lancet* 2000, 355, 1310-1314.
12. Wasserman AL, Thompson EI, Wilimas JA, Fairclough DL. The psychological status of survivors of childhood/adolescent Hodgkin's disease. *AJDC* 1987, 141, 626-631.
13. Evans SE, Radford M. Current lifestyle of young adults treated for cancer in childhood. *Arch Dis Child* 1995, 72, 423-426.
14. Byrne J, Fears TR, Steinhorn SC, Mulvihill JJ, Connelly RR, Austin DF, Holmes GF, Holmes FF, Latourette HB, Teta J, Strong LC, Myers MH. Marriage and divorce after childhood and adolescent cancer. *JAMA* 1989, 262, 2693-2699.
15. Green DM, Zevon MA, Hall B. Achievement of life goals by adult survivors of modern treatment for childhood cancer. *Cancer* 1991, 67, 206-213.
16. Haupt R, Fears TR, Robison LL, Mills JL, Nicholson HS, Zeltzer LK, Meadows AT, Byrne J. Educational attainment in long-term survivors of childhood acute lymphoblastic leukemia. *JAMA* 1994, 272, 1427-1432.
17. Hays DM, Landsverk J, Sallan SE, Hewett KD, Patenaude AF, Schoonover D, Zilber SL, Ruccione K, Siegel SE. Educational, occupational, and insurance status of childhood cancer survivors in their fourth and fifth decades of life. *J Clin Oncol* 1992, 10, 1397-1406.
18. Holmes GE, Baker A, Hassanein RS, Bovee EC, Mulvihill JJ, Myers MH, Holmes FF. The availability of insurance to long-time survivors of childhood cancer. *Cancer* 1986, 57, 190-193.
19. Jacobson Vann JC, Biddle AK, Daeschner CW, Chaffee S, Gold SH. Health insurance access to young adult survivors of childhood cancer in North Carolina. *Med Pediatr Oncol* 1995, 25, 389-395.
20. Kelaghan J, Myers MH, Mulvihill JJ, Byrne J, Connelly RR, Austin DF, Strong LC, Wister Meigs J, Latourette HB, Holmes GF, Holmes FF. Educational achievement of long-term survivors of childhood and adolescent cancer. *Med Pediatr Oncol* 1988, 16, 320-326.
21. Kingma A, Rammeloo LA, Der Does-van den Berg A, Rekers-Mombarg L, Postma A. Academic career after treatment for acute lymphoblastic leukaemia. *Arch Dis Child* 2000, 82, 353-357.

22. Makipernaa A. Long-term quality of life and psychosocial coping after treatment of solid tumours in childhood. A population-based study of 94 patients 11-28 years after their diagnosis. *Acta Paediatr Scand* 1989, 78, 728-735.
23. Meadows AT, McKee L, Kazak AE. Psychosocial status of young adult survivors of childhood cancer: a survey. *Med Pediatr Oncol* 1989, 17, 466-470.
24. Nicholson HS, Byrne J. Fertility and pregnancy after treatment for cancer during childhood or adolescence. *Cancer* 1993, 71, 3392-3399.
25. Nicholson HS, Mulvihill JJ, Byrne J. Late effects of therapy in adult survivors of osteosarcoma and Ewing's sarcoma. *Med Pediatr Oncol* 1992, 20, 6-12.
26. Novakovic B, Fears TR, Horowitz ME, Tucker MA, Wexler, LH. Late effects of therapy in survivors of Ewing's sarcoma family tumors. *J Pediatr Hematol Oncol* 1997, 19, 220-225.
27. Rauck AM, Green DM, Yasui Y, Mertens A, Robison LL. Marriage in the survivors of childhood cancer: a preliminary description from the Childhood Cancer Survivor Study. *Med Pediatr Oncol* 1999, 33, 60-63.
28. Teeter MA, Holmes GE, Holmes FF, Baker AB. Decisions about marriage and family among survivors of childhood cancer. *J Psychosoc Oncol* 1987, 5, 59-68.
29. Veenstra KM, Sprangers MA, van der Eyken JW, Taminiau AH. Quality of life in survivors with a Van Ness-Borggreve rotationplasty after bone tumour resection. *J Surg Oncol* 2000, 73, 192-197.
30. Zeltzer LK. Cancer in adolescents and young adults psychosocial aspects. Long-term survivors. *Cancer* 1993, 71, 3463-3468.
31. American Psychiatric Association. Diagnostic and statistical manual of mental disorders, 4th edition. 1994. Washington DC, American Psychiatric Association.
32. Butler R, Rizzi L, Handwerger B. The assessment of posttraumatic stress disorder in pediatric cancer patients and survivors. *J Pediatr Psychol* 1996, 21, 499-504.
33. Kazak AE, Barakat LP, Meeske K, Penati B, Barakat LP, Christakis D, Meadows AT, Casey R, Stuber ML. Posttraumatic stress, family functioning, and social support in survivors of childhood leukemia and their mothers and fathers. *J Consult Clin Psychol* 1997, 65, 120-129.
34. Stuber ML, Nader K, Yasuda P, Pynoos R, Cohen S. Stress responses after pediatric bone marrow transplantation: Preliminary results of a prospective longitudinal study. *J Am Academy Child Adolescent Psychiatry* 1991, 30, 952-957.
35. Hobbie WL, Stuber M, Meeske K, Wissler K, Rourke MT, Ruccione K, Hinkle A, Kazak AE. Symptoms of posttraumatic stress in young adult survivors of childhood cancer. *J Clin Oncol* 2000, 18, 4060-4066.
36. Survivors of Childhood Cancer: Assessment and Management. St Louis, Missouri: Mosby-Year Book, Inc 1994.
37. Nir Y. Post-traumatic stress disorder in children with cancer. In: Eth S, Pynoos R, editors. Posttraumatic Stress Disorder in Children. Washington, DC: American Psychiatric Press 1985, 123-132.
38. Pot-Mees C. The psychosocial effects of bone marrow transplantation in children. Thesis. Eburon Delft, The Netherlands, 1989.
39. Stuber M, Christakis D, Houskamp B, Kazak A. Post trauma symptoms in childhood leukemia survivors and their parents. *Psychosomatics* 1996, 37, 254-261.
40. Pelcovitz D, Goldenberg Libov B, Mandel F, Kaplan S, Weinblatt M, Septimus A. Posttraumatic stress disorder and family functioning in adolescent cancer. *J Traumatic Stress* 1998, 11, 205-221.
41. Stuber ML, Meeske K, Gonzalez S, Houskamp B, Pynoos R. Post-traumatic stress after childhood cancer I: the role of appraisal. *Psychooncology* 1994, 3, 305-312.
42. Greenberg HS, Kazak AE, Meadows AT. Psychologic functioning in 8- to 16-year-old cancer survivors and their parents. *J Pediatr* 1989, 114, 488-493.
43. Horowitz M, Wilner N, Alvarez W. Impact of Event Scale: a measure of subjective stress. *Psychosom Med* 1979, 41, 209-218.
44. Van der Velden P, Van der Burg S, Steinmetz CHD, Van der Bout J. Slachtoffers van bankovervallen (Victoms of bank robberies). Houten, The Netherlands: Bohn Stafleu Van Lochem 1992.
45. Schwardwald J, Solomon Z, Weisenberg M. Validation of the impact of event scale for psychological sequelae of combat. *J Consult Clin Psychooncol* 1987, 55, 251-256.

46. Epping-Jordan J, Compas B, Howell D. Predictors of cancer progression in young adult men and women: Avoidance, intrusive thoughts and psychological symptoms. *Health Psychol* 1994, 13, 539-547.
47. Cohen J. Statistical power analysis for the behavioral sciences. New York: Academic Press 1977.
48. Davidson JRT, Fairbank JA. The epidemiology of posttraumatic stress disorder. In: Davidson J, Foa E, editors. Posttraumatic stress disorder: DSM-IV and beyond. Washington, D.C. Amercian Psychiatric Press 1993, 147-169.
49. Kazak AE, Stuber ML, Barakat LP, Meeske K, Guthrie D, Meadows AT. Predicting posttraumatic stress symptoms in mothers and fathers of survivors of childhood cancers. *J Am Academy Child Adolescent Psychiatry* 1998, 37, 823-831.
50. Gijsbers van Wijk CMT, Kolk AM. Sex differences in physical symptoms: the contribution of symptom perception theory. *Soc Sci Med* 1997, 45, 231-246.
51. Phillips DL, Segal BE. Sexual status and psychiatric symptoms. *Am Soc Review* 1969, 34, 58-71.
52. Rourke MT, Stuber ML, Hobbie WL, Kazak AE. Posttraumatic stress disorder: understanding the psychosocial impact of surviving childhood cancer into young adulthood. *J Pediatr Oncol Nurs* 1999, 16, 126-135.
53. Yehuda R. Post-traumatic stress disorder. *N Eng J Med* 2002, 346, 108-114.

Chapter 9

General discussion

For health care professionals involved in the care of children with cancer and their families, it has been apparent for many years that not all the children who have reached the end of their treatment for their disease will have survived without adverse effects. For the child, at least three aspects of their lives will be affected in some way: physical, psychological and social [1]. Each of these aspects will together or separately influence the quality of survival for children and, when they grow up, possibly the quality of survival in the rest of their lives.

In this thesis we studied various aspects of quality of survival in young adult survivors of childhood cancer. A group of long-term survivors, aged 16 to 49 years old, were asked to fill in a questionnaire during their annual evaluation at the follow-up clinic. To put the results into perspective, a large group of young adults with no history of cancer was recruited with the help of survivors' general practitioners (GPs). The results of the separate studies described in the thesis are summarised in Table 1. In this last chapter, comments on study design, the most important results, the implications for clinical practice and directions for future research are discussed. Finally, the role of nurses in late effect evaluations is briefly discussed.

Comments on study design

In order to investigate various aspects of quality of survival, some choices were made concerning the selection of respondents in the described studies, as well as choices about the methods by which the different aspects could be assessed. The following section discusses these choices and their consequences.

Previous research on long-term survivors has often been conducted with relatively small samples, has examined a relatively short period after diagnosis, did not have comparisons with control groups and used unstandardised, study-specific instruments (*Chapter 3*). In the current studies, we attempted to overcome some of these problems by evaluating a large cohort of survivors who were an average of approximately 16 years since completion of treatment. Further, we selected a large control group and used mostly standardised instruments. Especially, the large cohort of survivors and the size of the reference group with almost equal aged peers is unique in the late-effects literature. However, some limitations of the findings need to be mentioned.

Study populations
The possibility of selection bias for both the survivors and the reference group can not be ruled out. With regard to the survivors, patients attending a long-term

follow-up clinic may not be fully representative of the larger population of survivors of childhood cancer. It is possible that survivors participating in the current study may have more somatic or psychological problems than non-compliant survivors, thereby biasing research findings. In our study group, we also have an over-representation of survivors with leukaemias and lymphomas. Due to logistic considerations, many of these survivors were seen in the follow-up clinic during the first years after the clinic was set up, which was also the time that we recruited survivors for our study.

For the reference group, we achieved, with the help of many GPs, a response rate of 63%, which is highly satisfactory for a mailed survey. However, we do not have data from non-participating persons. Although we asked all the eligible persons in the cover letter to let us know the reason why they refused to participate, only 24 complied with our request. As we asked the GPs to select persons and to send the questionnaires, we did not have names and addresses from the chosen persons. Moreover, we wrote in our cover letter to the controls that the GPs were the only ones who knew this data and that for us the persons stayed anonymous. We found it not ethical to ask the GPs for names and addresses to write to the persons ourselves, nor did we want to bother the GPs again by asking them to find out the reasons non-respondents had for not participating in the study. But because of this we have to take into account selection bias of the reference group. On the one hand, it is possible that persons who decided not to join the study are individuals with a lot of physical or psychological problems who did not have energy or interest to participate, which has probably introduced an underestimation of the level of QL. On the other hand, it is possible that exactly those persons with a lot of problems felt that they were called to participate.

Time since completion of therapy and type of cancer
Another limitation stems from one of the strengths of the study: the average time since completion of therapy. The survivors were diagnosed over a long period of time, namely 1963 to 1992. Impressive advances have been made during that period, not only in treatment but also in the field of supportive and psychosocial care for the patients. Special support programs have been developed over the years to improve care and minimise the adverse effects of illness and treatment. For instance, before 1978 there were not many facilities available for children to continue their education when they were admitted in the hospital and also skilled liaisons who helped families with the school system to get the best possible education were not present. This lack of attention to the child's education needs could have strongly affected the educational level of some survivors and perhaps even negatively influenced their QL.

The patients were also treated in a variety of ways. It was impossible to take into account all the different treatment regimens so we categorised treatment in a crude way in our studies. We have included survivors with different cancers, and did not take into account differences in severity of the cancer, intensity of treatment, initial prognosis, number of relapses or survivors' perceived treatment intensity and life threat.

Medical Limitation Scale

We used an adapted version of the Greenberg, Meadows & Kazak's Scale for Medical Limitations [2] to score survivors' late effects and health problems. Patients were categorised into the following three groups according to their most serious medical limitation: 1) mild = no limitations of activity; 2) moderate = no serious restriction of daily life; and 3) severe = significant restriction on daily activity or severe cosmetic changes (for a detailed description see Measures section in *Chapter 5, 6, 7, and 8*). Information regarding late effects/health problems was abstracted from the medical chart by one of the investigators (NL). The abstracted chart data were then rated by the same investigator, who is a paediatric oncology nurse and two paediatric oncologists. Patient names were excluded from the abstracted information to allow ratings to be blind to patient identity. As not all late effects/health problems are listed in one of the three categories, or as others are not defined in detail, it was not always easy to put a specific late effect/health problem in one of the groups. For instance, an absent limb or a second malignant neoplasm is a clear description, but a learning delay, a mild scoliosis or fatigue is not well defined and can lead to confusion. Therefore, some discrepancies in scoring were found between the different raters, however, all of these were resolved after a short or longer discussion.

There is no uniformity in the literature about the assessment of severity for late toxicity and several systems have been developed for reporting and grading late effects resulting from all modalities of treatment. In general, most systems are more or less similar to the grading system used by our team. Garré and colleagues [3] developed a set of criteria with toxicity grades including: grade 0) absent; grade 1) asymptomatic changes that did not influence the subject's general activity; grade 2) moderate symptomatic changes interfering with subject's activity; grade 3) severe symptomatic changes that required major corrective measures and strict and prolonged surveillance; and grade 4) life-threatening sequelae. The Swiss Pediatric Oncology Group (SPOG) developed a grading system for late effects in childhood cancer survivors and used it in 30 patients [4]. The system also ranges from 0 to 4, with grade 0) no late effect; grade 1) asymptomatic patient requiring no therapy; grade 2) asymptomatic patient, requires continuous therapy or continuous medical follow-up or symptomatic late effects resulting in reduced school, job, or psychosocial adjustment while remaining fully independent; grade 3) physical or mental sequelae not likely to be improved by therapy but able to work at least partially; and grade 4) severely handicapped patients unable to work independently. Two other rating scales focus specifically on the physical and cosmetic impairments. O'Malley and colleagues [5] developed the Combined Physical and Visible Impairment Rating Scale and Mulhern and colleagues [6] used an adapted version of the O'Malley questionnaire, the so-called Composite Cosmetic and Functional Impairment Ratings. Oeffinger and colleagues selected the second version of the Common Toxicity Criteria (CTCv2) in their study to grade late effects [7]. The CTC was created in 1988 by the National Cancer Institute (NCI) as an additional outcome measure to compare the acute toxicities of different treatments. In 1998, the NCI released the CVCv2, which included the Radiation Therapy Oncology Group/European Organization for Research and Treatment of Cancer Late Radiation Morbidity Scoring Scheme.

Table 1. Summary of the results of the studies presented in this thesis

Chapter	Characteristics of study sample	Purpose	Measures	Main results
2	Survivors: N=976 53% men / 47% women mean age at follow-up: 25 years mean age at diagnosis: 7 years mean time since diagnosis: 18 years mixed diagnoses	To describe medical and psychosocial problems encountered in a cohort of childhood cancer survivors seen at the follow-up clinic	Survivors were screened according to protocols based on the previously used treatment modalities	A total of 4004 medical or psychosocial problems was registered. The median number of problems found per survivor was 3. Almost a quarter of the survivors had maximally one problem registered. Survivors who have had a brain tumour or Ewing's sarcoma had the highest number of problems, survivors of Hodgkin's disease and acute lymphoblastic leukaemia the lowest
3	Studies: N= 30 well-matched control group or norm data available survivors primary source of information[a] survivors diagnosed before 20 years of age survivors at least 5 years after completion of therapy[b] survivors at least 18 years at the time of investigation[c]	To give an overview of the results of studies into the quality of life of young adult survivors of childhood cancer	A literature search of studies up to 2001 using the data bases of MEDLINE, CINAHL, EMBASE and PsychINFO	Although the literature yield some inconsistent findings, a number of trends could be identified: 1) most survivors reported to be in good health, with the exception of some bone tumour survivors; 2) most survivors function well psychologically; 3) survivors of CNS tumours and survivors of ALL are at risk for educational deficits; 4) job discrimination, difficulties in obtaining work and problems in obtaining health and life insurance's were reported; 5) survivors have lower rates of marriage and parenthood; 6) survivors worry about their reproductive capacity and/or about future health problems their children might experience as a result of their cancer history
4	Survivors: N= 35 29% men / 71% women mean age at follow-up: 27 years mean age at diagnosis: 7 years mean time since completion of therapy: 17 years mixed diagnoses	To explore the concept of fatigue from the survivors' perspective	A semi-structured interview with the following topics: description of fatigue and symptoms, frequency and course of fatigue, impact of fatigue on daily life, factors worsening or relieving fatigue, hours and pattern of sleep, and onset of fatigue	Most survivors diagnosed with cancer in their adolescence identified fatigue as a significant side-effect of the treatment. The majority of survivors who were toddlers or preschooler at the time of cancer treatment mentioned that, as far as they could recall, they had suffered from fatigue their entire life. The course of fatigue during the day differed among the survivors, although the majority reported to be fatigued when waking up in the morning. None of the survivors reported sleep problems. Many survivors slept 9 hours or more. Fatigue was defined by all respondents as having a negative impact on their daily lives
5	Survivors: N= 416 52% men / 48% women mean age at follow-up: 24 years mean age at diagnosis: 8 years mean time since completion of therapy: 15 years mixed diagnoses	What is the level of fatigue in comparison with a non-cancer group? What is the relation between demographic, medical and treatment characteristics and depressive symptoms and survivors' fatigue?	The Multidimensional Fatigue Inventory (MFI-20) The Center for Epidemiologic Studies Depression scale (CES-D)	Small differences were found in mean scores for the different dimensions of fatigue between survivors and controls. Survivors scored significantly lower (i.e., reflecting less fatigue) on general fatigue and reduced motivation, but statistically higher (i.e., reflecting worse fatigue) for mental fatigue than controls. Women experienced more fatigue than men. Linear regression revealed that female gender and

154

#	Demographics	Research question	Instruments	Results
	Controls: N= 1026 45% men / 55 % women mean age at follow-up: 26 years			unemployment were the only demographic characteristics explaining the various dimensions of fatigue. With regard to medical and treatment factors, diagnosis and severe late effects/health problems were associated with fatigue. Depression was significantly associated with fatigue on all subscales
6	Survivors: N= 400 55% men / 45 % women mean age at follow-up: 24 years mean age at diagnosis: 8 years mean time since completion of therapy: 16 years mixed diagnoses Controls: N= 560 45% men / 55% women mean age at follow-up: 26 years	What is the quality of life, the level of self-esteem and the degree of worries in comparison with a non-cancer group? What is the relationship between demographic, medical and treatment factors and self-esteem on the one hand and survivors' quality of life and degree of worries on the other hand?	The Medical Outcome Study Scale (MOS-24) The Worry questionnaire The Rosenberg Self-Esteem Scale (RSE)	Small differences were found in mean MOS-24 scores between survivors and controls. Survivors scored significantly lower levels of physical functioning, but statistically higher on vitality and general health perceptions than controls. No significant difference was found in the mean self-esteem scores. Female survivors had more cancer-specific worries than male survivors. In several related areas of general health, self-image and dying, survivors reported less worries than controls, but survivors worried significantly more about their fertility, getting/changing a job and obtaining insurance's. Female gender, age at follow-up, unemployment, lower educational level, years since completion of therapy, type of treatment, severe late effects/health problems and a low self-esteem were associated with a lower quality of life and a higher degree of concerns
7	Survivors: N = 500 53% men / 47% women mean age at follow-up: 24 years mean age at diagnosis: 8 years mean time since completion of therapy: 15 years mixed diagnoses Controls: N= 1092 45% men / 55% women mean age at follow-up: 26 years	What is the level of psychosocial adjustment with respect to educational status, employment status, living situation, marital status and offspring in comparison with a non-cancer group? What is the influence of demographic, medical and treatment factors on survivors' psychosocial adjustment?	Self-report questionnaire with three sections (with a total of 10 items): educational achievement, employment status, living situation, marital status and offspring	Many survivors were functioning well, however, a subgroup was less likely to complete high-school, to attain an advanced graduate degree, to follow normal elementary or secondary school and had to be enrolled more often on learning disabled programs. Less survivors were employed, but more survivors were student or homemaker. Survivors had lower rates of marriage and parenthood, worried more about their fertility and the risk of their children having cancer. Especially male survivors lived more often with their parents. Cranial irradiation dose ≤25 Gy was associated with a lower educational level. Brain/CNS tumour survivors had a higher risk of being single than survivors with a diagnosis of leukaemia/non-Hodgkin's lymphoma without cranial irradiation
8	Survivors: N = 500 53% men / 47% women mean age at follow-up: 24 years mean age at diagnosis: 8 years mean time since completion of therapy: 15 years mixed diagnoses	What is the level of posttraumatic stress symptoms? What is the relation between demographic, medical and treatment factors and survivors' posttraumatic stress symptoms?	The Impact of Event Scale (IES), a self-report instrument consisting of two subscales, Intrusion and Avoidance.	Twelve percent of survivors had scores in the severe range, indicating that these persons are unable to cope with the impact of their disease and need professional help. Twenty percent of the female survivors had scores in the severe range compared with 6% of the male survivors. Linear regression revealed that being female, unemployed, a lower educational level, diagnosis and severe late effects/health problems were associated with posttraumatic stress symptoms

[a] Studies with no more than 20% proxies as primary source of information were included as well
[b] Some studies included survivors who were less than 5 years after completion of therapy, these studies were included as well
[c] Some studies included survivors who were younger than 18 years old, these studies were included as well

Although the CTCv2 was not developed specifically for use in grading late effects, according to Oeffinger and colleagues the majority of late effects can be scored easily for the various organ systems or the appendixed Late Radiation Morbidity Scoring Scheme.

To our knowledge, none of the above described systems have yet been validated in the long-term cancer survivor population. It is needed that more homogeneous criteria of evaluation for the severity of late effects is defined and multi-institutional collaboration is recommended. It would be most helpful if a universally accepted, validated, scoring system would be available for use in grading late effects and that not everyone tries to re-invent the wheel.

Selection of quality of life measures
One of the difficulties in the interpretation of studies on QL in survivors of childhood cancer is the diversity of measures used (*Chapter 3*), and no consensus has yet been reached as to which are the best tools. As no standardised instrument for the quality of long-term survival related to long-term survivors of childhood cancer was available, we choose to administer a battery of tests to achieve comprehensiveness. The selected tests were mostly nationally and internationally widely used standardised psychometric instruments and measured the following topics: demographic and medical information, including treatment characteristics; QL comprising physical, psychological, sexual, and social functioning; experienced health status and worries about health; fatigue; self-esteem, depression, post-traumatic stress; educational achievement; occupational status; work and insurance discrimination; life style and philosophy of life and relationship with family and friends. To enhance the face validity of the data, the items were checked by medical, nursing and psychological staff experienced in the care of childhood cancer or in QL research. Additionally we piloted the questionnaire on 10 survivors before the study. The final questionnaire was identical both for the survivors and the comparison group except for a few questions that specifically addressed the issue of former illness and treatment. In these cases the term 'cancer' was replaced by 'any health problem'.

The overall study generated information that is not completely covered in this thesis. These include relationship with family and friends and life style and philosophy of life. We need to analyse this material further and it will hopefully be presented in future publications.

Results, implications for clinical practice and directions for future research

Recently, a number of studies which explored the overall morbidity within a population of young adult survivors of childhood cancer have been reported [3,4,7,8]. The percentage of childhood cancer survivors with a late effect diagnosed in the follow-up ranged from 58% to 71% in these studies. Multiple late effects were common, occurring in approximately 32-49% of survivors. Not all studies use the same definition of late effects and classification system and this makes comparison between the different studies virtually impossible. As said before, more homogeneous criteria of evaluation for the severity of late effects is needed. Although our study certainly has some limitations, as described in *Chapter 2*, the results showed that the majority of survivors seen at the follow-up clinic suffer from one or more medical and/or psychosocial problem. The continuously changing treatment options are expected to lead to differences between currently observed late treatment effects and the effects expected to be seen in the future. This fact strongly supports the need for continuous long-term follow-up of paediatric cancer survivors. Such follow-up would allow for continuous adjustments of the follow-up programmes based on relative risk estimates. The development of future follow-up programmes for childhood cancer survivors should, however, also be influenced by the impact or the quality of life of both the observed late effects and the screening programmes on the survivor. Another factor that must certainly be taken into account are the wishes of the childhood cancer survivors themselves that may also be expected to change over time.

Fatigue

In the study described in *Chapter 2*, a subgroup of childhood cancer survivors suffered from fatigue. The definition of the symptom fatigue can be challenged but in our follow-up clinic it denotes that fatigue 1) must be present for 6 months or more, 2) does not disappear after rest, and 3) interferes with daily life. Because no other studies where one scored fatigue in childhood cancer survivors as a late effect are available, it is difficult to compare our results with other findings. Most of what is known about off-treatment fatigue in childhood cancer survivors stems from 4 studies which have assessed fatigue in the context of a wider investigation on psychological adjustment or quality of life after cancer treatment [9-12]. In 2 of these studies the percentages of survivors suffering from fatigue were mentioned, namely 5% [9] and 8% [10], respectively. When we look at the percentages in adult cancer survivors suffering from off-treatment fatigue, the percentages are higher, namely 17% to 30% [13]. It is partly unknown why fatigue was not considered as a specific late effect in the above mentioned studies that assessed the overall morbidity in survivors. Oeffinger and colleagues [7] are the only ones who gave as a reason that subjective reports or symptoms that might overlap with other unrelated conditions, such as fatigue, dyspepsia, and insomnia, were not included as late effect. The other authors did not mention fatigue in their studies at all. What does this mean? Did the authors have the same arguments as Oeffinger and colleagues? Did survivors participating in those studies not suffer from fatigue? Did survivors not discuss fatigue with their physician during their visit at the follow-up clinic? Or was fatigue not recognised or acknowledged by health care providers? In a study done by Vogelzang and colleagues [14], the authors found that fatigue is seldom discussed by patients and their oncologists and that both have difficulty communicating about fatigue. Another reason for a possible underreporting might be the lack of energy to make clinicians understand symptoms or effects, particularly when they show little interest in or concern about the late effects or the survivor. In addition, the rush and excitement of the clinical environment may cause survivors to forget about significant issues they wanted to discuss until afterward when they are at home. Be that as it may, with so many survivors in our follow-up clinic who complained about fatigue, our interest in this phenomenon was awakened. After discussions with a number of survivors we suspected that we were dealing with a pathological entity related to the former cancer or its treatment, because the complaints were expressed with considerable uniformity. In these days, we did not know much about fatigue in survivors and information about off-treatment fatigue was very scarce in both adult and childhood cancer literature. As we felt that we needed more knowledge about fatigue to better assess survivors for the presence of fatigue and to work with them to develop a plan of care for its management, we decided to explore the experience of fatigue in the survivor group. An exploratory investigation was undertaken first, consisting of the qualitative study of fatigue as reported in *Chapter 4*. We used a qualitative approach because such methods are most suitable when the focus of a study is on particular experiences (in this case fatigue) where little is known [15]. As with quantitative research, the goals are to describe, explain, predict and control. However, this is accomplished not through establishing causality but through improving our comprehension of the phenomenon as a whole [16]. After a semi-structured interview in 35 survivors it was possible to identify some interesting key findings, namely that fatigue 1) is described as exerting oneself to the utmost, both physically and mentally; 2) is present most or all of the time since diagnosis and/or treatment; 3) is frequently already present upon awakening, despite an average of 9 hours sleep per night; and 4) negatively affects daily and social activities. Although our study certainly had some limitations, such as an over-representation of women in our study (71% women versus 29% men), and that we asked the survivors to describe in retrospect their experiences with fatigue, which might lead to recall bias, it gave us a better understanding of how the survivor experienced and expressed fatigue. It was obvious that fatigue was a serious problem for some young adult survivors and that many were limited in their ability to carry out their usual daily activities. The finding that many respondents indicated that fatigue developed shortly before their illness or during the treatment period supported our hypothesis that survivors' fatigue was probably related to their former illness or treatment.

The symptom of fatigue, however, is not specific for cancer. It is known that fatigue and lack of energy is a prevalent symptom in the general population [17,18] and also a major complaint among general practice attenders [19,20]. To interpret the significance of results obtained in follow-up studies involving childhood cancer survivors, a comparison should therefore be made with persons without a history of cancer. Therefore, we assessed the level of fatigue in a larger cohort of survivors and compared the results with a group of peers with no history of cancer (*Chapter 5*), and investigated a number of factors associated with survivors' fatigue. Contrary to our expectation, we could not demonstrate a significant difference in the level of fatigue between the survivors and controls. Survivors even reported less fatigue than their peers on the dimensions general fatigue and reduced motivation. Although previous studies which included a non-cancer comparison group also found no differences in fatigue scores [21,22], we, based on the findings of our qualitative study, more or less expected that survivors would experience more fatigue than the controls. As we described in our study the lack in difference in fatigue scores can be caused by a number of factors, such as the possible selection bias in the reference group, the influence of response shift, and the method of assessment. However, we still have no evidence that survivors' complaints of fatigue or lack of energy are characteristic for cancer survivors and it is clear that fatigue is not limited to cancer survivors alone. There is a risk that our findings are grist to the mill of health care professionals who remain sceptical about fatigue and think "it's all in the mind" or are convinced that all patients who complain about fatigue are depressed. The fact that in our study depression was strongly associated with fatigue could further support these thoughts. In view of traditional medical practice, which tends to emphasise objective clinical measures such as diagnostic tests and laboratory values, it is understandable that one finds it particularly frustrating to be confronted with a problem that cannot be touched or seen and for which there is no direct and objective evidence. But with an attitude of ignoring and rejecting fatigue in patients suffering from fatigue we fail in our duty towards our patients. Too many survivors who suffered from fatigue were ignored, mocked or accused of malingering in the past already [23]. Whatever the causes of fatigue in childhood cancer survivors are, whether it is mental, emotional, physical, or spiritual, survivors experiencing fatigue need help.

Where do we go from here?

The science of off-treatment fatigue is in its teenage years. Although there is nearly a century of physiological and psychological fatigue-oriented research that can be applied to understanding fatigue as experienced by people with cancer or to those who live with the effects of the treatment, there is still much to be done. From a clinical perspective it is of the utmost importance that health care providers recognise and acknowledge this symptom. They need to develop resources and skills to facilitate the assessment and management of fatigue. Fatigue should be incorporated in routine assessments of survivors who are being after completing treatment. The assessment will be most complete if it seeks details about both physical and mental fatigue. Some techniques such as checklists and short rating scales may improve clinical as well as research information gathering because they validate fatigue (i.e. "It is normal to acknowledge it, you don't have to be stoical'), and they give a frame of reference for comparison (more, less, a lot more, etc). They further allow questions about fatigue to be repeated at each visit and this is significant. Even though survivors' fatigue may not have changed, the way the survivor experience fatigue may well change [23]. If a survivor indicates symptoms of fatigue, a thorough assessment of all subjective and objective data that may influence fatigue for the survivor should be performed in order to design an effective plan of care for the survivor. It is recommended to involve partners, parents or other family members in this process, because in many cases family members may be more sensitive than the survivor to changes in the survivors' usual pattern and its impact. At the moment there is no "gold standard" for overall fatigue assessment. Recently several screening interviews and assessment guides have been suggested in the nursing literature [23] and recommendations vary according to the purpose of the assessment. In clinical practice, the focus must be on efficiently obtaining information that is needed for patient care. Some items for the assessment of fatigue, which is partly based on the work by Lenz and colleagues [24] are given in Appendix 1.

Once the evaluation is completed, identified energy leaks and other contributing factors can be addressed. Consultation from multidisciplinary team members in the development and implementation of the survivors' plan of care may be necessary. Psychologists and social workers can assist when fatigue impacts psychological or social well-being. Physicians, physiotherapists, and dieticians may be able to assist with the physical aspects of fatigue through pharmacological, exercise or diet interventions. Sometimes there is a great deal of ignorance among health care disciplines regarding the expertise that each of the other team members can bring to identify and solving the problem of fatigue creatively. One should realize that no single discipline will be able to find solutions for fatigue and that it is therefore essential to work together and to share insights and research findings with one another.

Fatigue interventions should be tailored to survivors' needs once the cause, intensity, and impact of fatigue has been determined. Many survivors need guidance from health care professionals in managing their fatigue. Patients tend to employ common-sense approaches and adopt measures that generally alleviate the normal tiredness that healthy people experience when lacking sleep or following exertion [25,26]. However, the passive approaches, like sleeping, resting and napping, frequently fail to alleviate the fatigue, and other self-care actions should be incorporated. Some clinical interventions commonly used for fatigue in patients are presented in Appendix 2, although most of these interventions are not evidence-based [27]. In addition, four other types of interventions for fatigue have been developed: exercise, preparatory education, attention-restoring activities, and psychosocial techniques [26]. Unfortunately, the benefits of these interventions are not always clear and not enough research has been conducted to suggest which interventions are the most viable and effective. It is also quite likely that, given the multicausal and multidimensional nature of fatigue, the use of single intervention strategies only provide partial relief. In a review of 22 studies, Trijsburg and colleagues [28] explored the effectiveness of psychological treatment for populations with cancer. This review concluded that "tailored counselling", where counselling and support were provided according to patients' needs, was effective not only for the reduction of distress and the enhancement of self-concept but also for the reduction of fatigue. This suggest that psychological support should form one aspect of a program for the management of fatigue. It goes without saying that this, as with other interventions, should be explicitly negotiated and agreed with the survivor. In our daily practice we sometimes see an unwillingness by the survivor to psychological approaches, perhaps from a concern that if the fatigue is labelled as psychological it implies fault. This erroneous belief deserves special attention.

It is obvious that we need to understand more about the nature and mechanisms of fatigue in childhood cancer survivors. There is a need for studies that use prospective longitudinal designs to yield more definitive information about the incidence and aetiology of fatigue in survivors and to identify those survivors who are most at risk. We also need more information about the outcome of fatigue. Impact on quality of life, mobility, self-care, social isolation, and role change need to be explored in more detail. Scientific inquiry should also direct future clinical interventions for treatment and assist individuals in coping with the experience of fatigue. Little is known how fatigue in patients with cancer differ from fatigue in other patient populations such as those with renal disease or chronic fatigue syndrome. In the year 2000 the Wilhelmina Kinderziekenhuis in Utrecht and the Emma Kinderziekenhuis AMC started a pilot study in which off-treatment fatigue in adolescents with cancer is compared with fatigue in adolescents with chronic fatigue syndrome. The first results of this study are not expected before 2003-2004. Further, before persons become a survivor they once were a child or adolescent in treatment for cancer. Special attention must be paid on the incidence and causes of fatigue in this population.

Quality of life, psychosocial functioning and posttraumatic stress symptoms
While much of the focus of care during active treatment has been the support of physical and psychological well-being, concerns about late effects and its treatment on physical and psychological well-being most often arise several months to years later [29]. These late effects, while not life-threatening, can affect day to day functioning.
Overall, the results of the studies in this thesis are in agreement with studies that conclude that long-term survivors of childhood cancer are, for the most part, well-adjusted, functioning well and leading normal lives [30-33]. However, we also found that some survivors do experience negative adverse effects which may cause little or major interruptions in their lives. The findings suggest the following.

Quality of life
1) The health-related QL of survivors, as measured with the MOS-24, is not different from their peers. Female gender, unemployment, a lower educational level, type of treatment and severe late effects/health problems, and a low self-esteem were identified as predictors for a poorer QL in survivors.
2) Although many survivors worried not more or even less about several issues than their peers, they often are more concerned about some present and future concerns, such as fertility, getting a job, and obtaining insurance's. When one looks at the cancer-specific concerns in detail, 54% of the survivors worried about having a relapse, 50% expressed their concern about having another cancer when they are older, and 43% worried about the health of their future children. Female gender, unemployment and a low self-esteem were associated with a higher degree of worries.

Psychosocial functioning
1) Educational achievement is affected in a subgroup of survivors. Especially female survivors were less likely to complete high-school or to attain an advanced graduate degree in comparison with their female peers. Cranial irradiation was the strongest independent prognostic factor of educational achievement.
2) Some survivors reported to have experienced some form of job discrimination as a result of their health history.
3) Survivors, especially males, tend to live more often with their parents than their peers and were less frequently married or lived together than controls.

A history of a brain/CNS tumour was identified as a risk factor for being single.
4) The percentage of survivors with biologic children was significantly lower than in the comparison group.

Posttraumatic stress symptoms
1) A substantial group of childhood cancer survivors report symptoms of posttraumatic stress. Female gender, unemployment, a lower educational level, type of diagnosis, and severe late effects/health problems were associated with posttraumatic stress symptoms. However, these factors could explain stress levels only to a limited extent.

These findings suggest a number of implications for health care professionals who work with childhood cancer survivors and can help them to target their efforts toward those survivors most likely to need their services.

Overall, the health-related QL of survivors, as measured with the MOS-24, was not different from their peers. In fact, survivors reported even a better QL on several dimensions than controls. In *Chapter 6*, we discussed potential explanations for the finding of the lack of difference in scores, such as resilience, response-shift, denial, and a bravado coping style. It is also possible that the instrument we used was unsuitable for measuring health-related QL in survivors. Although generic instruments measure health functioning across a wide variety of diseases and provide a common data base for comparing results, allocating resources, and developing health policy [34], they may not identify issues unique to the cancer experience. The vast majority of standardised QL tools focus primarily on acute treatment effects, and often do not contain a longer view of cancer survivorship with specific concerns and long term needs [35]. The development of a specific instrument for childhood cancer survivors may overcome this problems. An assessment can than be targeted to the former disease, the age group of the persons under review, a certain function, or to a particular problem (such as mobility following major limb surgery). This will allow us to pick up finer gradations between categories and to explore other domains. Unless an accurate, disease-specific sensitive measure is developed to identify the issues that face young childhood cancer survivors so that problems can be addressed, the real outcome in terms of the impact of therapy will be partly unknown and it will be impossible to compare one treatment regimen with another.

We tried to identify those survivors who are most at risk to have a poorer QL and female gender, unemployment, a lower educational level, severe late effects/ health problems, and a low self-esteem were identified as predictors. Special attention has to be paid to these survivors because they may be in need of more help and support. Specific strategies and approaches to assess and support self-esteem in survivors as means of enhancing QL, such as support groups or coping skills, need to be explored and evaluated. However, it should be noticed that the explained variance by these variables was small. Further investigation on potential determinants of QL in survivors is necessary.

Worries

Our results show that survivors often worry about having a relapse, their fertility, about the health of their future children, and about having another cancer when they are older. Concerns about relapse are an almost universal response in the first months and years after active treatment ends. When times passes by and all goes well, these fears will normally fade away and will no longer be a part of their daily, weekly, or monthly reality. It is known that many survivors feel anxious when it is time for their annual follow-up appointment, afraid that "they will find out that something is wrong". These feelings are very common and most survivors has his or her own way to deal with this. However, some survivors continue to have deep fears over an extended period of time and have, even many years after treatment, nightmares or anxiety attacks that may interfere with their daily life. If these fears grows to large, it may sometimes compromise the survivors' ability to seek appropriate health care and some of them will stay away from the follow-up clinic. Little information exists on how survivors express fear of recurrence or what coping mechanisms they use to control their stress.

A big concern for many survivors is whether they will be able to have children. Although the majority of survivors remain fertile after their treatment, fertility is affected in some survivors. In the past it happened that some survivors were never told or, and this still happens today, did not recall hearing that infertility is a potential consequence of treatment for their cancer. Other survivors do not learn that their ability to have children may have been compromised or destroyed until they have spent several emotional and expensive years trying [36]. Survivors could benefit greatly from being informed early about possible sterility so that they can have time to investigate and plan for alternative ways of becoming a parent.

Survivors also often worry about the health of their future children. They sometimes wonder if they could pass on their cancer genetically to their children, or they are afraid that a child conceived after treatment might be born defective or disabled. The results of studies looking at the rate of birth defects in children born to childhood cancer survivors are very encouraging. In general, children born to survivors are just as healthy as those born to people who never had cancer [37]. However, a very few types of cancer can be passed from parent to child genetically, such as retinoblastoma. A truthful and realistic picture should be given to the survivor, so that he or she is aware of possible problems and knows about the help that can be provided. Survivors with a higher risk for having a child with health problems might want to consider genetic counselling and possibly genetic testing.

Finally, some survivors worry that treatment or genetic predisposition or the combination of both will result in a second cancer. Although for most of the survivors the chance of getting a second cancer is very small, some survivors indeed have an increased risk of developing a second malignancy. While survivors cannot alter a genetic predisposition or undo the damage from radiation, they can modify their risk factors by performing health behaviour. Therefore, survivors should be encouraged to follow the recommendations for risk reduction and early detection of cancer, including no smoking, the use of sunscreens and simple screening methods such as breast self-examination.

Educational achievement

Our study *(Chapter 7)* showed an impairment in educational achievement in survivors. Cranial irradiation was an important explanatory factor of lower educational level. This finding is not new. It is known for years that disease and treatment can have potential deleterious effects on future intellectual functioning, especially in children with leukaemias, lymphomas and brain tumours. To date, most clinics have skilled school liaisons who can help children and families work with the school system to get the best possible education. It is important, however, that care and support for these children does not stop when treatment ends and the child is no longer admitted in the hospital. Survivors who are at risk for learning difficulties may benefit from ongoing assessment of school performance and/or neuropsychological testing. If educational needs can be identified and appropriate interventions initiated as soon as possible, it may be possible to minimise the child's academic problems. Recent research shows that children with treatment-related learning problems may be helped by cognitive rehabilitation therapy [38], which involves multiple training sessions with education therapists who focus on improving memory, attention, and math skills. Children are taught strategies they can use to improve their ability to learn. When the child grows older, career and employment plans should be discussed.

Discrimination

In the past, the diagnosis of cancer was presumed to mean inevitable and relatively rapid death [39]. As a result, individuals cured of cancer often carried the stigma of survival, were regarded as different and were discriminated in education, employment, or health, life, and disability insurance coverage [40]. In 1987, van Eys [41] wrote "that as long as survivors are viewed as persons rescued from the dead, strangers in a strange land, they may be perceived with open arms but, by virtue of their special status, excluded from society".

The Nederlandse Kankerbestrijding and other groups devoted to the goal of informing the public about cancer have done much to change the negative impact of a cancer diagnosis. Considering the findings in our study where some survivors reported to have experienced some form of job discrimination as a result of their health history, this remains a important task. We also tried to obtain information about health and life insurance's from survivors in our study (data not presented in this thesis). It became clear that many survivors were very ill-informed about their insurance's and could not answer our questions, resulting in unreliable data. However, from clinical reports of survivors we know that many of them are often confronted with refusals, restricted policies, and higher premiums. For these reasons those who work with cancer survivors must formally and informally endeavour to educate the uninformed general public. Speaking to interested community groups, publishing the positive facts of cancer treatment as well as the negative ones, and recognising the success in cancer treatment, are ways health care professionals can shape more positive attitudes.

Living situation, marriage and offspring
The findings in our study suggest that survivors, and especially males, tend to live more often with their parents than their peers (*Chapter 7*). Although there is a tendency for young adults in the Netherlands to stay longer with their parents and live on one's own at an older age, there is a serious discrepancy between the survivors and the comparison group. In our study we also asked respondents some questions regarding overprotection by parents (not covered in this thesis). The results showed that, compared with the controls, both male and female survivors felt significantly more that their parents worried too much about them and treated them still as a child. Many survivors also reported that their parents were overprotective and reluctant to let them go. These findings may indicate that the illness experience amongst these survivors has hindered development towards normal independence in relation to their families. It is known, that due to the former illness, an unusually strong bond tends to develop between the parents and child, sometimes to the extent that they depend exclusively on one another for the satisfaction of emotional and physical needs. This bond may jeopardize the normal process of development of autonomy and the way to independence might be delayed. More extensive exploration of this issue is needed.

The finding that, compared with controls, survivors were less likely to marry or lived together and that a history of a brain/CNS tumour is a risk factor for being single is in agreement with results in previous studies [31,42,43]. Several suggestions has been made in the literature regarding the causes of the marriage deficit, such as substantial problems in the areas of growth and motor, and difficulties in social and emotional development [42]. The marriage deficit among survivors of brain/CNS tumours may be related to the long-term sequelae that are unique to this group (including decreased intelligence and neuropsychologic and memory problems) [43] and perhaps to the societal standard that holds males responsible for support of their spouses. In the study by Green and colleagues [31], almost 16% of the survivors who were never married or had lived as married reported that their history of childhood cancer had influenced their decisions regarding marriage. Koocher and O'Malley reported that issues that influenced the decision to marry included concerns of the patient about the effect of treatment on sexual function, and concerns of the patient's spouse regarding the potential for having healthy children [44]. Perhaps it is also possible that survivors simply delay marriage for a longer time than members of the general population. Research that pinpoint the reasons for not marrying or living together is recommended. This latter recommendation goes also for studies about the offspring of survivors. It speaks more or less for itself that as survivors tend to be less often married or living together, a lower percentage of survivors will have children. The factor that some survivors are infertile or have an impaired fertility from their treatments may play a role as well.

Posttraumatic stress symptoms

"Although I am more than fifteen years past treatment, I still can't enter a hospital without experiencing strong physical reaction to my memories of cancer treatment.Whenever I enter an oncology clinic, I break into a cold sweat and my heart starts racing. The sight of alcohol pads and the sight of butterfly IVs never fail to make me nauseous"[36] (page 46)

Although supportive of the occurrence of posttraumatic stress symptoms in young adult survivors of childhood cancer, our data described in *Chapter 8* are from a preliminary investigation. Two limitations need to be mentioned. First, we did not use structured diagnostic interviews but found our findings through a self-report questionnaire. One has to be careful to base a diagnosis of posttraumatic stress that rests on self-report symptoms. Second, little is known about the validity of the Impact of Event Scale and if there is an association with clinical diagnosis of PTSD based on diagnostic measures. However, we know from clinical practice that a subgroup of survivors is likely to have re-experiencing, arousal, and avoidant symptoms that revolve in large part around issues of health care and/or illness. Events such as driving to the hospital, or smells associated with their treatment may be reminders potent enough to generate strong physical and emotional responses. It is possible that such symptoms may have an influence of survivors' health care behaviours. For example, survivors may excessively attend to and overinterpret physical symptoms and seek medical care often in unnecessary situations. Alternatively, and perhaps of greater concern, survivors may avoid seeking medical care because it is a powerful and painful reminder of their experiences, which causes them to stay away from important follow-up visits or even avoid seeking treatment when ill [45].

The strongest take-home message from the data is that additional research is urgently needed and that posttraumatic stress should be considered by health care providers that escort childhood cancer patients in their rehabilitation back to normal life.

The role of nurses in late effect evaluations
Although it is not entirely within the scope of our study, the last part of this chapter discusses the role of nurses in late effect evaluations and some suggestions are made.

As it became apparent that survivors often had multiple health care and psychosocial needs, some paediatric oncologists started follow-up clinics to provide a multidisciplinary team that monitor and support survivors. These follow-up clinics not only provide care for survivors, but also participate in research projects that track the effectiveness and side effects of various treatments. Ideally the multidisciplinary team should consist of paediatric oncologists, internist-oncologists, radiation oncologists, nurses and various psychosocial personnel including a psychologist, social worker, and a school liaison. In addition, physicians from related disciplines such as cardiology, endocrinology, orthopedics, and neurology must be associated with the team to assist in the evaluation of identified problems [46]. However, follow-up programmes varies widely within the United States [47], and, although there is no data available, probably in Europe and the rest of the world. On the whole, there is little consensus on how follow-up should take place and how the care must be provided.

Follow-up clinics have been in existence in the United States since 1983; the first was established at the Children's Cancer Research Center of Children's Hospital of Philadelphia. The focus for follow-up was to provide comprehensive and systematic evaluation and treatment and to 1) decrease the negative impact of long-lasting effects of therapy, 2) assist the survivor and their families to cope effectively while monitoring and treating late effects, and 3) help the survivor and their families gain perspective on the cancer experience so they can be vigilant toward potential late effects [48]. A paediatric oncology nurse was recruited to identify eligible patients from their population to inform them of the services available and to begin to meet their needs. The role of this nurse was found to be multifaceted and consisted of the functions of clinician, caretaker, educator, and researcher [49]. This was the beginning of many follow-up clinics in the United States, where a paediatric nurse practitioner, specialised in oncology, fulfils the role of clinic co-ordinator. Over time, their role has evolved in a fashion consistent with the trends described for advanced clinical nursing in general [46]. The primary focus of the role flows from the direct care of individual survivors and high-risk subgroups to indirect care of the clinic population of survivors as a collective and of the community-at-large [46]. Now the role includes five functions: 1) specialty care provider, 2) educator, 3) researcher, 4) clinical/programme manager and, 5) consultant.

In general, the role of nurses in follow-up clinics in the United States is very clear and is one potential way of maximising the nursing contribution to follow-up. However, the role of the nurse practitioner in the Netherlands is less well defined than in the United States and if follow-up clinics and institutions are prepared to include a nurse practitioner to their team is doubtful. It remains also a debatable question if a nurse-led clinic is necessary and possible in the Netherlands at the moment. However, there are other possible roles the nurse could fulfil in a multi-disciplinary team.

It seems quite likely that there will be some development and changes in the structure of follow-up clinics in the Netherlands in the coming years in order to provide appropriate care to the increasing number of survivors. It is critical to recognise that follow-up of survivors requires a multidisciplinary approach, in the same way that treatment in the paediatric oncology wards brings together all of the relevant disciplines. As survivors have multiple needs and require intervention from several health professionals, it is essential that one is responsive to survivors' needs and that one make efforts to ensure that the service offered fulfils survivors' expectations. It is hoped that before one reorganises the follow-up clinic one takes a close look at all different roles and that roles will be evaluated and described. An evaluation of the current service could result in changes that is based on survivor and family need and not on tradition [50]. There exists both opportunity and challenge for paediatric oncology nurses at present to develop their role in the follow-up clinic.

Appendix 1.
Recommendations for the assessment of fatigue (based on the work by Lenz and colleagues) [24]

- Evaluate intensity and quality. (What is the strength or severity of the fatigue? Intensity can be quantified by asking the patient to rate his or her fatigue using a numeric intensity scale from 1 (no fatigue) to 10 (severe fatigue). What words does the patient use to describe the quality and nature of fatigue?)
- Evaluate onset, duration, and pattern. (When did the fatigue begin? What is the pattern? Does the fatigue fluctuate over time?)
- Evaluate meaning and level of perceived distress. (What meaning does the patient place on the fatigue? How much distress is this causing the patient? Why does the patient think that he or she is fatigued? What is the meaning of the impact of fatigue?)
- Evaluate physiological, psychological, and situational influencing factors. (The causes of fatigue are still not well understood. Several theoretical models on fatigue have proposed a variety of causes or influencing factors. The physiological, psychological, and situational factors described in this framework may give rise to or affect the experience of the symptom of fatigue. Physiological parameters may include type and stage of cancer, current and past treatment, other concurrent illnesses, the usage of medications, alcohol, or drugs, weight change, and associated symptoms (e.g. pain, nausea, sleep disturbance). Assessment of changes in haematological and biochemical status provide additional information on physiological status. Psychological factors encompass the mental state or mood (e.g. anxiety, depression or sadness, anger), changes in attitude and motivation, and changes in ability to concentrate and focus attention. Exploration of situational factors involves lifestyle, (nutrition, exercise, sleep/rest patterns), social support, culture, living arrangements, and physical environment)
- Evaluate impact of fatigue. (The assessment of the consequences of fatigue includes changes in physical and social activity pattern, activities of daily living, and role performance)
- Evaluate self-care. (Patients may initiate a range of self-care behaviours to relieve their fatigue. A review of the effectiveness of these behaviours for a patient may be helpful in the development of a management plan)

Appendix 2.
Interventions commonly used for patients with cancer

- Educate patients and families regarding deleterious effects of prolonged bedrest and too much inactivity
- Educate patients and families about expectancies of fatigue related to the disease and treatment
- Help patients and families identify what fatigue-promoting activities they can modify and how to modify them
- Encourage patients to maintain a journal to identify fatigue patterns
- Establish priorities for activities based on usual social role and cultural values
- Encourage activity within individual limitations: make goals realistic
- Suggest individualised environmental or activity changes that may off-set fatigue
- Maintain adequate hydration and nutrition
- Monitor the effects of fatigue on quality of life
- Evaluate the efficacy of self-care fatigue interventions on a regular and systematic basis
- Promote an adequate balance between activity and rest
- Recommend physical therapy referral for patients with specific neuro-musculoskeletal deficits
- Schedule important daily activities during times of least fatigue and eliminate nonessential, unsatisfying activities
- Address the negative impact of psychological and social stressors and how to modify or avoid them

References

1. Bradwell M, Hawkins J. Survivorship and rehabilitation. In: Langton H, editor. The Child with Cancer-Family Centered Care in Practice. London: Bailliere Tindall 2000.
2. Greenberg HS, Kazak AE, Meadows AT. Psychologic functioning in 8- to 16-year-old cancer survivors and their parents. *J Pediatr* 1989, 114, 488-493.
3. Garre ML, Gandus S, Cesana B et al. Health status of long-term survivors after cancer in childhood. Results of an Uniinstitutional study in Italy. *Am J Pediatr Hematol Oncol* 1994, 16, 143-152.
4. von der Weid N, Beck D, Caflisch U, Feldges A, Wyss M, Wagner HP. Standardized assessment of late effects in long-term survivors of childhood cancer in Switzerland: results of a Swiss Pediatric Oncology Group (SPOG) pilot study. *Int J Pediatr Hematol Oncol* 1996, 3, 483-490.
5. O'Malley JE, Foster D, Koocher G, Slavin L. Visible physical impairment and psychological adjustment among pediatric cancer survivors. *Am J Psychiatr* 1980, 137, 94-96.
6. Mulhern RK, Wasserman AL, Friedman AG, Fairclough D. Social competence and behavioral adjustment of children who are long-term survivors of cancer. *Pediatrics* 1989, 83, 18-25.
7. Oeffinger KC, Eshelman DA, Tomlinson GE, Buchanan GR, Foster BM. Grading of late effects in young adult survivors of childhood cancer followed in an ambulatory adult setting. *Cancer* 2000, 88, 1687-1695.
8. Stevens MC, Mahler H, Parkes S. The health status of adult survivors of cancer in childhood. *Eur J Cancer* 1998, 34, 694-698.
9. Wasserman AL, Thompson EI, Wilimas JA, Fairclough DL. The psychological status of survivors of childhood/adolescent Hodgkin's disease. *AJDC* 1987, 141, 626-631.
10. Kanabar DJ, Attard-Montalto S, Saha V, Kingston JE, Malpas JE, Eden OB. Quality of life in survivors of childhood cancer after megatherapy with autologous bone marrow rescue. *Ped Hemat Oncol* 1995, 12, 29-36.
11. Moe PJ, Holen A, Glomstein A, Madsen B, Hellebostad M, Stokland T, Wefring KW, Steen-Johnson J, Nielsen B, Howlid H, Borsting S, Hapnes C. Long-term survival and quality of life in patients treated with a national all protocol 15-20 years earlier: IDM/HDM and late effects? *Ped Hemat Oncol* 1997, 14, 513-524.
12. Zeltzer LK, Chen E, Weiss R, Guo MD, Robison LL, Meadows AT, Mills JL, Nicholson HS, Byrne J. Comparison of psychologic outcome in adult survivors of childhood acute lymphoblastic leukemia versus sibling controls: a cooperative Children's Cancer Group and National Institutes of Health study. *J Clin Oncol* 1997, 15, 547-556.
13. Servaes P, Verhagen C, Bleijenberg G. Fatigue in cancer patients during and after treatment: prevalence, correlates and interventions. *Eur J Cancer* 2002, 38, 27-43.
14. Vogelzang NJ, Breitbart W, Cella D, Curt GA, Groopman JE, Horning SJ, Itri LM, Johnson DH, Scherr SL, Portenoy RK. Patient, caregiver, and oncologist perceptions of cancer-related fatigue: results of a tripart assessment survey. *Semin Hematol* 1997, 34, 4-12.
15. Morse JM. Qualitative Nursing Research. A Contemporary Dialogue. Newbury Park, California: Sage Publications 1991.
16. Burns N, Grove SK. Introduction to qualitative research. In: Kay D, editor. The practice of nursing research: conduct, critique and utilization. Philadelphia: W.B. Saunders Company 1987, 75-106.
17. Chen MK. The epidemiology of self-perceived fatigue among adults. *Preventive Med* 1986, 15, 74-81.
18. Lewis G, Wessely S. The epidemiology of fatigue: more questions than answers. *J of Epidem Comm Health* 1992, 46, 92-97.
19. Bates DW, Schmitt W, Buchwald D, Ware NC, Lee J, Thoyer et al. Prevalence of fatigue and chronic fatigue syndrome in a primary care practice. *Arch Intern Med* 1993, 153, 2759-2765.
20. Fuhrer R, Wessely S. The epidemiology of fatigue and depression: A French primary-care study. *Psychol Med* 1995, 25, 895-905.
21. Smets EMA, Visser MRM, Willems-Groot AFMN, Garssen B, Schuster-Uitterhoeve ALJ, de Haes JCJM. Fatigue and radiotherapy:(B) experience in patients 9 months following treatment. *Br J Cancer* 1998, 78, 907-912.
22. Hann DM, Jacobsen P, Martin S, Azzarello L, Greenberg H. Fatigue and quality of life following radiotherapy for breast cancer: a comparitive study. *J Clin Psychiatry in Med Sett* 1998, 1, 19-33.

23. Winningham ML, Barton-Burke M. Fatigue in cancer. A multidimensional approach. Sudbury, Massachusetts: Jones and Bartlett Publishers 2000.
24. Lenz ER, Pugh LC, Milligan RA, Gift A, Suppe F. The middle-range theory of unpleasant symptoms: an update. *Adv Nurs Sci* 1997, 19, 14-27.
25. Langeveld NE, Ubbink MC, Smets EMA. 'I don't have any energy': The experience of fatigue in young adult survivors of childhood cancer. *EJON* 2000, 4, 20-28.
26. Ream E, Richardson A. From theory to practice: designing interventions to reduce fatigue in patients with cancer. *Oncol Nurs Forum* 1999, 26, 1295-1305.
27. Winningham ML, Nail LM, Barton-Burke M et al. Fatigue and the cancer experience: the state of the knowledge. *Oncol Nurs Forum* 1994, 21, 23-36.
28. Trijsburg R, van Knippenberg F, Rijpma S. Effects of psychological treatment on cancer patients: a critical review. *Psychosom Med* 1992, 54, 489-517.
29. Ferrell BR, Hassey Dow K, Grant M. Measurement of the quality of life in cancer survivors. *Quality Life Res* 1995, 4, 523-531.
30. Gray RE, Doan BD, Shermer P, FitzGerald AV, Berry MP, Jenkin D, Doherty MA. Psychologic adaptation of survivors of childhood cancer. *Cancer* 1992, 70, 2713-2721.
31. Green DM, Zevon MA, Hall B. Achievement of life goals by adult survivors of modern treatment for childhood cancer. *Cancer* 1991, 67, 206-213.
32. Kazak AE. Implications of survival: Pediatric Oncology Patients and their Families. In: Bearison A, Mulhern R, editors. Pediatric Psychooncology. New York: Oxford University Press 1994, 171-192.
33. Meadows AT, McKee L, Kazak AE. Psychosocial status of young adult survivors of childhood cancer: a survey. *Med Pediatr Oncol* 1989, 17, 466-470.
34. Aaronson NK. Quality of life research in cancer clinical trials: A need for common rules and language. *Oncology* 1990, 4, 59-66.
35. Padilla GV, Grant M, Ferrell BR. Nursing research into quality of life. *Quality of Life* 1992, 1, 41-348.
36. Keene N, Hobbie W, Ruccione K. Childhood Cancer Survivors. A practical guide to your future. 1st ed. Sebastopol, CA: O'Reilly & Associates 2000.
37. Sankila R, Olsen JH, Anderson H, Garwicz S, Glattre E, Hertz H, Langmark F, Lanning M, Moller T, Tulinius H. Risk of cancer among offspring of childhood-cancer survivors. Association of the Nordic Cancer Registries and the Nordic Society of Paediatric Haematology and Oncology. *N Eng J Med* 1998, 338, 1339-1344.
38. Hendriks CMCM, van den Broek-Sandmann TM. Amsterdamse Training van Aandacht en Geheugen voor Kinderen (ATAG-K). Swets & Zeitlinger, the Netherlands 1996.
39. Cunnick WR, Cromie JB, Cortell RE, Wright BP. Employing the cancer patient: A mutual responsibility. *J Occupat Med* 1974, 16, 775-780.
40. Hobbie W, Ruccione K, Moore IK, Truesdell S. Late effects in long-term survivors. In: Foley GV, Fochtman D, Mooney KH, editors. Nursing Care of the Child with Cancer. Orlando, Florida: W.B. Saunders Company 1993, 466-496.
41. van Eys J. Living beyond cure: Transcending survival. *Am J Pediatr Hemat Oncol* 1987, 9, 114-118.
42. Byrne J, Fears TR, Steinhorn SC, Mulvihill JJ, Connelly RR, Austin DF, Holmes GF, Holmes FF, Latourette HB, Teta J, Strong LC, Myers MH. Marriage and divorce after childhood and adolescent cancer. *JAMA* 1989, 262, 2693-2699.
43. Rauck AM, Green DM, Yasui Y, Mertens A, Robison LL. Marriage in the survivors of childhood cancer: a preliminary description from the Childhood Cancer Survivor Study. *Med Pediatr Oncol* 1999, 33, 60-63.
44. Koocher GP, O'Malley JE. The Damocles syndrome: Psychosocial consequences of surviving childhood cancer. New York: McGraw-Hill 1981.
45. Rourke MT, Stuber ML, Hobbie WL, Kazak AE. Posttraumatic stress disorder: understanding the psychosocial impact of surviving childhood cancer into young adulthood. *J Pediatr Oncol Nurs* 1999, 16, 126-135.
46. Survivors of Childhood Cancer: Assessment and Management. St Louis, Missouri: Mosby-Year Book, Inc 1994.
47. Oeffinger KC, Eshelman DA, Tomlinson GE, Buchanan GR. Programs for adult survivors of childhood cancer. *J Clin Oncol* 1998, 16, 2864-2867.

48. Hollen PJ, Hobbie W. Establishing comprehensivespecialty follow-up clinics for long-term survivors of cancer. *Support Care Cancer* 1995, 3, 40-44.
49. Hobbie W. The role of the pediatric oncology nurse specialist in a follow-up clinic for long-term survivors of childhood cancer. *J Pediatr Oncol Nurs* 1986, 3, 13-21.
50. Gibson F, Soanes L. Long-term follow-up following childhood cancer: maximising the contribution from nursing. *Eur J Cancer* 2001, 37, 1859-1868.

Summary

Cancer among children is relatively uncommon, with approximately 400 children 0 to 14 years of age being newly diagnosed with cancer each year in the Netherlands. Since the 1970s, survival rates for children have significantly improved and nowadays, the cure rate is about 70%. As survival has greatly improved, the need to assess the quality of survival has increased correspondingly. In the past decade it has become evident that former anti-cancer treatment can cause harmful side effects in many survivors, some of them developing many years after treatment. These side effects include both medical and psychosocial problems that may affect the survivors' physical, mental or social health.

The main subject of this thesis was to study the quality of survival in young adult survivors of childhood cancer. Therefore, a cohort of 500 long-term survivors, aged 16 to 49 years old, were asked to fill in a questionnaire during their annual evaluation at the follow-up clinic. The different aspects of survivors' health outcomes was compared with that of a group young adults with no history of cancer.

Chapter 1 gives a general overview of the most important aspects of childhood cancer. Incidence, survival rates and treatment modalities are described. The somatic early and late side effects of both the disease and the therapy are described, together with the impact of the treatment period on the child. Further, the need for follow-up in this patient group is discussed and a brief description of the follow-up outpatient clinic in The Emma Kinderziekenhuis/Academic Medical Center is given. Finally, the concept of quality of life (QL) is defined.

Chapter 2 reports on the medical and psychosocial problems encountered temporarily or permanently in a cohort of 976 childhood cancer survivors who were evaluated at the long-term follow-up clinic in our hospital. Survivors were screened by a paediatric oncologist (persons aged <18 years) or medical oncologist (persons aged >18 years) for late medical effects. For psychosocial effects they were seen by a research nurse or psychologist. Screening for late effects was performed according to protocols based on the previously used treatment modalities. Although the study certainly has some limitations, the results showed that the majority of survivors suffer from one or more medical and/or psychosocial problems. A total of 4004 medical or psychosocial problems were registered for the 976 survivors. The median number of problems found per survivor was three. Almost a quarter of the survivors had maximally one problem registered. Survivors who have had a brain tumour or Ewing's sarcoma had the highest number of problems, survivors of Hodgkin's disease and acute lymphoblastic leukaemia the lowest. The continuously changing treatment options are expected to lead to differences between currently observed late treatment effects and the effects expected to be seen in the future. In our opinion, this fact strongly supports the need for continuous long-term follow-up of paediatric cancer patients.

Chapter 3 presents a review of the research on QL in young adult survivors of childhood cancer. A search was made in MEDLINE, CINAHL, EMBASE and Psychinfo for studies up to 2001. The review has been performed according to the following methodological criteria: 1) well-validated and reliable measures, 2) well-matched control group, or comparison with norms, 3) information about demographic, medical and treatment factors, 4) respondent rate, 5) use of appropriate statistical tests, 6) survivors as the primary source of information, age at diagnosis before 20 years of age, and 8) at least 5 years after completion of therapy. Thirty studies were found that met the inclusion criteria, the majority of studies were conducted in the United States. Studies were characterised by a high degree of heterogeneity with respect to: the patients sample employed (e.g. survivors with different cancers who had undergone a variety of treatments), the comparison groups selected, the QL dimensions assessed and the instruments employed. Age at time of evaluation, age at diagnosis, and time elapsed since completion of therapy varied widely. Further, the majority of the studies reviewed suffered from at least one of the following methodological weaknesses: the use of small samples, unstandardised, study-specific instruments, and cross-sectional rather than prospective designs.

Despite the heterogeneity in study procedures and the methodological shortcomings, a number of clear trends could be identified: 1) most survivors reported to be in good health, with the exception of some bone tumour survivors; 2) some studies mention fatigue as a residual effect of treatment; 3) the majority of survivors function well psychologically. Female gender, older age at follow-up, greater number of relapses, presence of severe functional impairment, and cranial irradiation were associated with an increased risk for emotional problems in some studies; 4) survivors of CNS tumours and survivors of ALL are at risk for educational deficits. Cranial irradiation and an early age at diagnosis was associated with educational deficits; 5) job discrimination, difficulties in obtaining work and problems in obtaining health and life insurance's were reported; 6) survivors seem to stay at home longer after reaching adulthood and leave home at an older age in comparison to their peers; 7) survivors have lower rates of marriage and parenthood; 8) survivors worry about their reproductive capacity and/or about future health problems their children might experience as a result of their cancer history.

In our opinion, there is a need for methodologically studies that measure QL among survivors of childhood cancer more precisely by taking into account the effects of the severity of the cancer and the long-term impact of different treatments. Further, additional data are needed to understand the needs of survivors and to identify those subgroups of survivors who are at greatest risk to the adverse sequelae of the disease and its treatment.

In *Chapter 4*, a qualitative study is presented in which 35 long-term survivors were interviewed in depth about their fatigue. A subgroup of survivors complained about extreme fatigue during their annual follow-up visit. Although it is speculated that fatigue occurs equally in adults, children and adolescents with cancer, little research exists to substantiate this view. Due to the current lack of knowledge, more information on the phenomenology of fatigue of childhood cancer survivors is desirable.

The aim of the study was to explore the concept of fatigue from a survivors' perspective. Survivors were regarded as suffering from fatigue if they indicated that 1) fatigue was present for 6 months or more, 2) not disappeared after rest, and 3) interfered with their daily life. The topics which were covered during the interview included the nature, onset and pattern of fatigue, sleep rest pattern, what helps with fatigue and what does not help, and the impact of fatigue on their daily life.

The mean age of the survivors was 27 years (range 18-38 years) and 71% were women. Fifteen persons were treated for leukaemia or lymphoma and 20 persons for a solid tumour. The average time since the end of treatment was 17 years (range 8-25 years). Results indicate that most survivors who were diagnosed with cancer in their adolescence identified fatigue as a significant side effect of the treatment. The majority of survivors who were toddler or pre-schooler at the time of cancer treatment mentioned that, as far as they could recall, they had suffered from fatigue their entire life. The course of fatigue during the day differed among the survivors, although the majority reported to be fatigued when waking up in the morning. None of the survivors reported sleep problems. A striking finding was the amount of sleep time reported by the survivors. About half of the survivors slept between 7.5 and 9 hours per night, others slept more than 9 hours. A variety of activities that helped to decrease fatigue were described, such as sleep, rest and recreational activities. Fatigue was defined by al respondents as having a negative impact on their daily lives. For some survivors the impact of fatigue extended to the work setting, they often described having less energy to continue their work.

We conclude that fatigue is a serious problem for some young adult survivors of childhood cancer. Although the findings of qualitative studies are not meant to be generalized, our findings should be kept in mind by nurses, physicians and other health-care providers in oncology follow-up clinics. As some survivors are now recognised to possibly be at risk of fatigue, we recommend the routine screening of all survivors for this symptom. It is also important to investigate the prevalence of fatigue in a comparable young general population sample to ascertain whether survivors of childhood cancer differ, as far as fatigue is concerned, from healthy persons.

Chapter 5 presents a study where we assessed the level of fatigue in 416 long-term survivors (age range 16-49 years, 48% female) and 1026 peers with no history of cancer (age range 16-53 years, 55% female). Demographic, medical and treatment characteristics associated with survivors' fatigue were identified and the association between depressive symptoms and survivors' fatigue was studied. Fatigue was measured with the Multidimensional Fatigue Inventory (MFI-20), a self-report instrument consisting of five subscales (general fatigue, physical fatigue, mental fatigue, reduced activity, and reduced motivation). Depression was measured with a part of the Center for Epidemiologic Studies Depression scale (CES-D). Multiple linear regression analysis identified the independent exploratory variables of the five fatigue dimensions.

Small differences were found in mean scores for the different dimensions of fatigue between survivors and controls (range effect sizes -0.34 to 0.34). Survivors scored significantly lower (i.e., reflecting less fatigue) on general fatigue and reduced motivation, but statistically higher (i.e., reflecting worse fatigue) for mental fatigue than controls. Women experienced more fatigue than men. Multiple linear regression revealed that being female and unemployed were the only demographic characteristics explaining the various dimensions of survivors' fatigue. With regard to medical and treatment factors, type of diagnosis and severe late effects/health problems were associated with survivors' fatigue. Finally, depression was significantly associated with survivors' fatigue on all subscales. The selected characteristics explained only a moderate proportion of the variability (R^2) of the fatigue scores: 29%-46%.

We conclude that although our clinical practice suggests a difference in fatigue between survivors and their peers, this could not be confirmed in our study using the MFI-20. The well known correlates between depression and fatigue was confirmed in our study. Further research is needed to clarify the undoubtedly complex somatic and psychological mechanisms for the development, maintenance and treatment of fatigue in childhood cancer survivors.

In *Chapter 6*, the QL, self-esteem and worries in survivors was evaluated and compared with that in a group of young adults with no history of cancer. In addition, the impact of demographic, medical and treatment features and self-esteem on survivors' QL and worries was studied. Respondents self-assessment of QL was measured by the Medical Outcome Study Scale (MOS-24), a 24 item short form health survey consisting of seven dimensions: physical functioning, role functioning, social functioning, mental health, vitality, bodily pain, and general health perceptions. To investigate the respondent's concerns, we used the Worry questionnaire which includes three subscales: cancer-specific concerns, general health concerns, and present and future concerns. Finally, self-esteem was assessed with the Rosenberg Self-Esteem questionnaire. Multiple linear regression analysis identified the independent explaining variables of the QL dimensions and worries.

A total of 400 survivors (age range 16-49 years, 45% female) and 560 controls (age range 16-53 years, 55% female) completed the questionnaires. Small to moderate differences were found in mean MOS-24 scores between survivors and controls (range effect sizes -0.36-0.22). Survivors scored significantly lower levels of physical functioning, but statistically higher on vitality and general health perceptions than controls. No significant difference was found in the mean self-esteem scores between survivors and controls. Female survivors had more cancer-specific concerns than male survivors. In several related areas of general health, self-image and dying, survivors reported less worries than controls, but survivors worried significantly more about their fertility, getting or changing a job and obtaining insurance's. Female gender, unemployment, severe late effects/health problems and a low self-esteem were independently related to poorer quality of life in survivors, whereas age at follow-up, unemployment, years since completion of therapy and a low self-esteem was associated with a higher degree of survivors' worries.

We conclude that the QL and the level of self-esteem in survivors of childhood cancer is not different from their peers. Although many survivors worried not more or even less about issues than their peers, they often are concerned about their own health and some present and future concerns. The investigated factors could explain worries and a poor QL only to a limited extent (8% for cancer-specific concerns to 37% for mental health). Further research exploring determinants and indices of QL and worries in survivors is warranted. From a clinical perspective, health care providers can use this knowledge to plan interventions to enhance the QL in survivors and to decrease the degree of worries.

In *Chapter 7*, we evaluated educational achievement, employment status, living situation, marital status and offspring in 500 survivors (age range 16-49 years, 47% female). The results were compared with a reference group of 1092 persons with no history of cancer (age range 15-33 years, 55% female). The contribution of various factors on survivors' psychosocial functioning was studied with multivariate logistic regression analysis. All participants completed a self-report questionnaire.

The results showed that, although many survivors were functioning well, a subgroup of survivors was less likely to complete high-school, to attain an advanced graduate degree, to follow normal elementary or secondary school and had to be enrolled more often on learning disabled programs. The percentage of employed survivors was significantly lower compared with the comparison group, but more survivors were student or homemaker. Survivors also experienced some form of job discrimination as a result of their health history. More than half of the male survivors were rejected for military service. Survivors, especially male survivors, lived more often with their parents and were less frequently married or lived together than controls. The survivor group also reported worrying more about their fertility and the risk of their children having cancer. The results of the logistic regression models showed that female survivors were more likely to have a lower educational level than male survivors. In contrast, male survivors were more likely to live with their parents and to be single. Survivors with a history of brain/CNS tumours had a higher risk of being single than survivors with a diagnosis of leukaemia/non-Hodgkin lymphoma without cranial irradiation. Finally, cranial irradiation was the strongest independent prognostic factor of educational achievement. Survivors who received a radiation dose of ≤25 Gy were about 8 times more likely to have a lower educational level than survivors without this type of treatment.

We conclude that, although many young adult survivors of childhood cancer function as well as their peers and seem to live a normal life with activities common for their age, a substantial number of survivors continue to experience residual limitations in their psychosocial functioning many years after their initial diagnosis and successful medical treatment. Follow-up of survivors is considered essential. It is important to investigate risk factors and causes of all possible adverse outcomes of childhood cancer and its treatment. Moreover, early identification and appropriate intervention may be in the interest of the individual survivor. Counselling and advice should be available.

In *Chapter 8*, we investigated posttraumatic stress symptoms in survivors using the Impact of Event Scale. In addition, the relationship of survivors' demographic, medical and treatment characteristics on posttraumatic stress symptoms was analysed. Results showed that 12% of this sample of adult survivors had scores in the severe range, indicating that these persons are unable to cope with the impact of their disease and need professional help. Twenty percent of the female survivors had scores in the severe range as compared with 6% of the male survivors. Linear regression revealed that being female, unemployed, a lower educational level, type of diagnosis and severe late effects/health problems were associated with a higher level of stress symptoms.

This finding support the outcomes reported previously that diagnosis and treatment for childhood cancer may have significant long-term effects, which are manifest in symptoms of posttraumatic stress. However, our findings should be viewed as preliminary. There is a need for psychiatric diagnostic interviews and further research exploring symptoms of posttraumatic stress in childhood cancer survivors in more detail is clearly warranted. From a clinical perspective, health care providers must pay attention to these symptoms during evaluations in the follow-up clinic. Early identification of posttraumatic stress symptoms can enhance the QL for survivors of childhood cancer.

In *Chapter 9*, the overall findings are discussed. In addition, the clinical implications for clinical practice are formulated and directions for future research are given. Finally, the role of nurses in late effect evaluations is briefly discussed.

Samenvatting

Kanker is een zeldzame ziekte op de kinderleeftijd, die elk jaar in Nederland bij ongeveer 400 kinderen in de leeftijd van 0 tot 14 jaar wordt vastgesteld. Sinds de jaren zeventig van de vorige eeuw, zijn de resultaten van de behandeling van kinderen met kanker aanmerkelijk verbeterd en hiermee de kans op genezing. Het genezingspercentage voor de hele groep patiënten is momenteel ongeveer 70%. Door de toegenomen kans op genezing wordt de behoefte om de kwaliteit van overleven te meten ook in toenemende mate van belang. In de loop der jaren is namelijk gebleken dat onderdelen van de behandeling, die personen met kanker krijgen, bij sommige patiënten schadelijke gevolgen hebben. Sommige gevolgen treden vrij snel op tijdens of na de behandeling, anderen kunnen pas jaren na het beëindigen van de behandeling duidelijk worden, de zogenaamde late effecten. Deze late effecten kunnen zowel somatische als psychosociale problemen veroorzaken die van invloed kunnen zijn op de kwaliteit van leven van de overlever (survivor).

Dit proefschrift richt zich voornamelijk op een aantal aspecten van kwaliteit van leven bij jong volwassenen die genezen zijn van jeugdkanker. Tijdens hun jaarlijkse bezoek aan de Polikliniek Late Effecten Kindertumoren (PLEK) werd aan 500 survivors, in de leeftijd van 16 tot 49 jaar, gevraagd een vragenlijst in te vullen. De resultaten werden vergeleken met een controlegroep van personen die in het verleden geen kanker hebben gehad.

In *Hoofdstuk 1* wordt een beknopt overzicht gegeven van de belangrijkste kenmerken van kanker bij kinderen. De incidentie, de genezingskansen en de behandeling van jeugdkanker worden beschreven. Hierna volgt een overzicht van de belangrijkste vroege en late somatische gevolgen die op kunnen treden als gevolg van de behandeling met chirurgie, chemotherapie en radiotherapie. Er wordt ingegaan op de periode van ziekzijn en behandeling en de mogelijke invloed daarvan op het verdere leven van het kind. De noodzaak van follow-up voor deze groep patiënten wordt belicht en er wordt kort ingegaan op het ontstaan van de PLEK in het Emma Kinderziekenhuis AMC. Als laatste wordt het begrip kwaliteit van leven gedefinieerd.

In *Hoofdstuk 2* worden de somatische en psychosociale late gevolgen beschreven die gevonden zijn bij 976 survivors die de PLEK hebben bezocht. De survivors zijn gescreend aan de hand van tevoren opgestelde protocollen door een kinderarts-oncoloog (leeftijd < 18 jaar) of een internist-oncoloog (leeftijd > 18 jaar) op lichamelijke gevolgen. De psychosociale gevolgen werden in kaart gebracht door een gespecialiseerde verpleegkundige of psycholoog. Hoewel het onderzoek een aantal beperkingen heeft, komt uit de gegevens naar voren dat het merendeel van de survivors een of andere vorm van lichamelijke of psychosociale schade heeft opgelopen door de vroegere behandeling. Totaal werden 4004 medische of psychosociale problemen geregistreerd bij de 976 survivors. De mediaan van het aantal problemen per survivor was drie. Bij een kwart van de survivors werd maximaal één probleem gevonden. Survivors die in het verleden een hersentumor of een Ewing sarcoom hebben gehad blijken het grootste aantal problemen te hebben, terwijl survivors die in het verleden Morbus Hodgkin of een acute lymfatische leukemie hebben gehad het laagste aantal problemen hebben. Hoogstwaarschijnlijk zullen er verschillen optreden tussen de huidige late gevolgen en de gevolgen die in de toekomst gevonden zullen worden door de steeds veranderende behandelingsmogelijkheden. Daarom is het noodzakelijk om kinderen die behandeld zijn voor kanker op lange termijn te blijven vervolgen.

Een overzicht van de literatuur op het gebied van kwaliteit van leven bij jong volwassenen die in hun jeugd behandeld zijn voor kanker wordt gepresenteerd in *Hoofdstuk 3*. Artikelen zijn gezocht met behulp van een bepaalde zoekstrategie in MEDLINE, CINAHL, EMBASE en Psychinfo over de periode 1966 tot 2001. Dertig studies die aan de inclusiecriteria voldeden werden in deze review opgenomen, de meeste studies uitgevoerd in de Verenigde Staten. Onderzoeken verschilden aanzienlijk van elkaar met betrekking tot de onderzochte patiëntengroepen (b.v. survivors met verschillende maligne aandoeningen die allerlei behandelingen hadden ondergaan), de geselecteerde controlegroepen, de gemeten kwaliteit van leven dimensies en de gebruikte meetinstrumenten. Daarnaast verschilden de leeftijd op het moment van evaluatie, de leeftijd tijdens diagnose, en de jaren na het einde van de behandeling sterk. Verder bleken de onderzoeken belangrijke methodologische beperkingen te hebben zoals kleine onderzoeksgroepen en niet-gestandaardiseerde, studie-specifieke instrumenten.
Ondanks deze verschillen in methode en methodologische beperkingen konden een aantal trends worden waargenomen: 1) de meeste survivors gaven aan een goede gezondheid te hebben, met uitzondering van enkele survivors die behandeld waren voor een bottumor; 2) vermoeidheid werd genoemd als laat gevolg in een aantal studies; 3) de meeste survivors functioneerden goed op psychologisch gebied. Het vrouwelijk geslacht, oudere leeftijd tijdens evaluatie, het aantal recidieven, aanwezigheid van ernstige lichamelijke beperkingen en schedelbestraling bleken geassocieerd te zijn met een minder goed psychologisch functioneren; 4) survivors met tumoren van het centraal zenuwstelsel en survivors met acute lymfatische leukemie hebben een verhoogd risico op leerproblemen. Schedelbestraling en een jonge leeftijd tijdens diagnose waren geassocieerd met leerproblemen; 5) discriminatie op het werk, problemen in het vinden van een baan en het verkrijgen van ziekte- en levensverzekeringen werden door de survivors genoemd als probleem; 6) survivors lijken langer bij hun ouders te blijven wonen en op latere leeftijd het huis uit te gaan in vergelijking met hun leeftijdsgenoten; 7) het percentage getrouwde survivors is lager, ook hebben ze minder vaak kinderen; 8) survivors maken zich zorgen over hun vruchtbaarheid en/of hun kinderen gezondheids-problemen zullen krijgen doordat zij vroeger kanker hebben gehad.
We concluderen dat er behoefte is aan methodologische studies die meer rekening houden met de ernst van de doorgemaakte maligniteit en de invloed van de

verschillende behandelingen op de kwaliteit van leven van de survivor. Bovendien is er meer inzicht nodig in de behoeften die de survivors hebben ten aanzien van hun kwaliteit van leven, als mede het identificeren van survivors of survivor groepen die een risico lopen op late somatische en psychosociale gevolgen van ziekte en behandeling.

In *Hoofdstuk 4* wordt een kwalitatieve studie beschreven waarin 35 survivors werden geïnterviewd over hun vermoeidheid. Tijdens hun eerste bezoek aan de PLEK gaven veel survivors aan ernstig vermoeid te zijn. Alhoewel uit onderzoek gebleken is dat vermoeidheid bij volwassen kankerpatiënten vaak voorkomt, is er bij personen die jeugdkanker hebben gehad nog nauwelijks onderzoek gedaan naar deze klacht. Gezien dit gebrek aan kennis is het wenselijk dat er meer informatie komt over het vóórkomen en de verschijnselen van vermoeidheid in deze groep. Het doel van dit onderzoek was dan ook inzicht te krijgen in de vermoeidheid zoals die ervaren werd door personen die in hun jeugd behandeld zijn voor kanker. Vermoeidheid werd in dit onderzoek gedefinieerd als vermoeidheid welke 1) minimaal 6 maanden aanwezig was, 2) niet verdween na rust, en 3) de dagelijkse activiteiten deed afnemen. Tijdens het interview kwamen de volgende onderwerpen aan bod: de aanvang, duur, intensiteit en het dagelijkse patroon van de vermoeidheid, het slaappatroon, factoren die de vermoeidheid verergeren en verminderen, en de invloed van de vermoeidheid op het dagelijkse leven. De gemiddelde leeftijd van de survivors was 27 jaar (range 18-38 jaar) en het percentage vrouwen was 71%. Vijftien personen zijn behandeld voor leukemie of een lymfoom en 20 personen voor een solide tumor. Het gemiddelde interval tussen het einde van de behandeling en de afname van het interview was 17 jaar (range 8-25 jaar). De resultaten gaven aan dat vermoeidheid bij de meeste personen die behandeld waren voor kanker tijdens de adolescentie al aanwezig was tijdens hun behandeling. Veel personen die behandeld werden toen zij peuter of kleuter waren rapporteerden dat zij, voor zover zij zich konden herinneren, hun hele leven al moe waren. Het dagelijkse patroon van de vermoeidheid bleek te verschillen onder de respondenten, alhoewel bijna iedereen aangaf al moe te zijn bij het wakker worden. Niemand rapporteerde slaapstoornissen. Verrassend was het aantal uren slaap gerapporteerd door deze jong volwassenen. Ongeveer de helft van de personen sliep 7.5 uur per nacht, de andere helft sliep meer dan 9 uur per nacht. Factoren die de vermoeidheid verminderden waren o.a. rust, slapen en afleiding. Iedereen gaf aan dat vermoeidheid hen in het dagelijks functioneren belemmerde en een negatieve invloed had op de kwaliteit van het leven. Deelname aan het arbeidsproces bleek voor een aantal personen problematisch door het afgenomen energie niveau.

We concluderen dat vermoeidheid een groot probleem is voor een aantal jong volwassenen die in hun jeugd behandeld zijn voor kanker. Het is van belang dat verpleegkundigen, artsen en andere hulpverleners, werkzaam op kinderoncologische follow-up klinieken dit probleem onderkennen. Door het stellen van een aantal korte vragen tijdens de jaarlijkse controle kan men inzicht krijgen in hoeverre dit probleem actueel is. Een andere aanbeveling met betrekking tot de onderzoekspopulatie is om een controlegroep van leeftijdsgenoten in het onderzoek te betrekken en hiervan doen we verslag in het volgende hoofdstuk.

Hoofdstuk 5 beschrijft de studie die vermoeidheid meet bij 416 survivors (range leeftijd 16-49 jaar, 48% vrouw) en 1026 leeftijdsgenoten die in het verleden geen kanker hebben gehad (range leeftijd 16-53 jaar, 55% vrouw). Tevens wordt, met behulp van multivariate lineaire regressie, de invloed van demografische en medische karakteristieken en depressie op de vermoeidheid van de survivor bestudeerd. Vermoeidheid werd gemeten met de Multidimensionele Vermoeidheids Index (MVI), een zelfrapportage instrument die de dimensies algemene vermoeidheid, lichamelijke vermoeidheid, mentale vermoeidheid, reductie in activiteit en reductie in motivatie bevat. Depressie werd gemeten met de Center for Epidemiologic Studies Depression Scale (CES-D).

Er werden kleine verschillen gevonden in de gemiddelde scores op de verschillende dimensies van vermoeidheid tussen de beide onderzoeksgroepen (range effect size -0.34-0.34). Vrouwen bleken meer vermoeid te zijn dan mannen. Logistische regressie toonde aan dat zowel het vrouwelijke geslacht als werkeloosheid de voornaamste demografische kenmerken zijn van de verschillende dimensies van vermoeidheid bij de survivor. Met betrekking tot de medische karakteristieken waren diagnose en ernstige late gevolgen/gezondheidsproblemen onafhankelijk geassocieerd met vermoeidheid bij de survivor. Naast de demografische en medische karakteristieken verklaarde depressie een deel van de verschillende dimensies van vermoeidheid. De verklaarde varianties van de geselecteerde karakteristieken varieerden van 29% (mentale vermoeidheid) tot 46% (algemene vermoeidheid).

We concluderen dat, in tegenstelling tot onze verwachting en dagelijkse praktijk, er geen grote verschillen lijken te bestaan tussen de beide onderzoeksgroepen wat betreft vermoeidheid. Het is bekend dat vermoeidheid en depressie met elkaar correleren, dit wordt in ons onderzoek bevestigd.

In *Hoofdstuk 6* worden kwaliteit van leven, zelfwaardering en zorgen geëvalueerd bij de survivors en de resultaten worden vergeleken met de referentiegroep. Vervolgens wordt de invloed van demografische en medische karakteristieken en zelfwaardering op de kwaliteit van leven en zorgen van de survivor onderzocht. De ervaren kwaliteit van leven werd gemeten met de MOS-24, een generieke lijst met 24 vragen en 7 dimensies: lichamelijk functioneren, rolvervulling, sociaal functioneren, psychische gezondheid, vitaliteit, lichamelijke pijn, en ervaren gezondheid. Met behulp van een zorgenvragenlijst, welke 3 subschalen bevat (kanker-specifieke zorgen, algemene gezondheidszorgen, huidige en toekomstige zorgen), werden de zorgen van de respondenten bestudeerd. Als laatste werd zelfwaardering gemeten met de Rosenberg Self-Esteem vragenlijst. De diverse associaties werden onderzocht met multivariate lineaire regressie modellen.

Vierhonderd survivors (range leeftijd 16-49 jaar, 45% vrouw) en 560 controlepersonen (range leeftijd 16-53 jaar, 55% vrouw) vulden de vragenlijsten in. Er werd weinig verschil gevonden tussen de gemiddelde scores van de beide onderzoeksgroepen op de subschalen van de MOS-24 (range effect sizes -0.36-0.22). Ook was er geen significant verschil in de gemiddelde zelfwaarderingsscore tussen survivors en controlepersonen. Vrouwelijke survivors gaven meer kankerspecifieke zorgen aan dan mannelijke survivors. Survivors rapporteerden minder zorgen dan controlepersonen op het gebied van algemene gezondheid, zelfbeeld en doodgaan, maar daarentegen hadden zij significant meer zorgen over hun vruchtbaarheid, een baan krijgen of van baan veranderen en het krijgen van een levens- of ziekteverzekering. Het vrouwelijk geslacht, werkeloosheid, ernstige late gevolgen/gezondheids-problemen en een lage zelfwaardering waren onafhankelijk geassocieerd met een aangetaste levenskwaliteit van de survivor, terwijl werkeloosheid, leeftijd tijdens diagnose, jaren na het stoppen van de behandeling en een lage zelfwaardering geassocieerd waren met meer zorgen van de survivor. De totale variantie van de MOS-24 en zorgen die verklaard konden worden door de onderzochte variabelen varieerde van 8% (kanker-specifieke zorgen) tot 37% (vitaliteit).

We concluderen dat de ervaren kwaliteit van leven en de zelfwaardering van de survivors niet veel verschilt van de ervaren kwaliteit van leven en zelfwaardering van hun leeftijdsgenoten. Alhoewel de meeste survivors zich niet meer of zelfs minder zorgen maken dan hun leeftijdsgenoten, zijn ze wel meer bezorgd over hun eigen gezondheid en een aantal huidige en toekomstige zorgen.

De geselecteerde karakteristieken verklaren slechts in beperkte mate een verminderde kwaliteit van leven en meer zorgen bij de survivors.

Hoofdstuk 7 beschrijft een studie waarin we opleiding, werk, woon- en leefomstandigheden en eigen kinderen bestuderen in 500 survivors (range leeftijd 16-49 jaar, 47% vrouw). De resultaten worden ook hier weer vergeleken met een referentiegroep van 1092 personen die geen kanker hebben gehad (range leeftijd 15-33 jaar, 55% vrouw). Daarbij werd nagegaan welke factoren een lage opleiding, wonen bij ouders en niet getrouwd zijn bij de survivors konden verklaren. We bestudeerden zowel demografische kenmerken (geslacht, leeftijd tijdens evaluatie) als medische kenmerken (leeftijd tijdens diagnose, jaren na het beëindigen van de behandeling, duur van behandeling, diagnose en behandeling met of zonder schedelbestraling).

De resultaten lieten zien dat, alhoewel veel survivors goed functioneerden, een subgroep van de survivors minder opgeleid was of een HBO of universitaire studie deed in vergelijking met hun leeftijdsgenoten. Ook waren meer survivors verwezen naar speciaal basisonderwijs. Het percentage survivors met een betaalde baan was significant lager dan het percentage controlepersonen met een betaalde baan, maar meer survivors waren nog bezig met een studie of werkten als huisvrouw. Survivors gaven aan dat zij op de een of andere manier gediscrimineerd werden op het werk. Meer dan de helft van de mannelijke survivors was afgewezen voor militaire dienst. Survivors, en vooral mannen, woonden vaker bij hun ouders en waren minder vaak getrouwd of samenwonend in vergelijking met hun leeftijdsgenoten. De survivors bleken verder meer zorgen te hebben over hun vruchtbaarheid en of hun kinderen kanker zouden krijgen dan de controlepersonen. Onafhankelijke voorspellende factoren voor een lage opleiding waren het vrouwelijk geslacht en behandeling met schedelbestraling. Survivors die behandeld waren met een bestralingsdosis van 25 Gy of minder hadden ongeveer 8 keer meer kans op een lagere opleiding dan survivors die op een andere manier behandeld waren. Onafhankelijke voorspellende factoren voor het wonen bij ouders waren het mannelijke geslacht, jongere leeftijd tijdens evaluatie en jonger dan 6 jaar tijdens diagnose. Tenslotte waren de onafhankelijke voorspellende factoren voor niet getrouwd zijn het mannelijke geslacht, jonge leeftijd tijdens evaluatie, jonger dan 6 jaar tijdens diagnose en diagnose. Survivors die behandeld waren voor een hersentumor of een tumor van het centraal zenuwstelsel hadden 4.5 keer meer kans om niet getrouwd te zijn dan survivors die behandeld waren voor een leukemie of een non-Hodgkin lymfoom zonder schedelbestraling.

Onze conclusie is dat, alhoewel een grote groep survivors net zo goed functioneert als hun leeftijdsgenoten, er een groep survivors is die een aantal beperkingen heeft vele jaren na hun ziekte en behandeling. Vanuit een klinisch perspectief is follow-up aan te bevelen.

In *Hoofdstuk 8* wordt een studie beschreven waarin posttraumatische stress symptomen bij de survivors wordt onderzocht. Daarnaast wordt de invloed van demografische en medische kenmerken op posttraumatische stress symptomen bij de survivor onderzocht. Uit het onderzoek kwam naar voren dat 12% van de survivors last had van ernstige posttraumatische stress symptomen. Vrouwen hadden meer last van ernstige posttraumatische stress symptomen dan mannen, respectievelijk 12% versus 6%. Het vrouwelijk geslacht, werkeloosheid, een lagere opleiding, diagnose en ernstige late gevolgen/gezondheidsproblemen bleken significant geassocieerd te zijn met posttraumatische stress symptomen. De resultaten lijken overeen te komen met bevindingen uit eerder onderzoek waarin men aangeeft dat bij kinderen die kanker hebben gehad en hiervoor succesvol behandeld zijn, angst- en vermijdingssymptomen voorkomen. Het blijkt dat deze symptomen het best begrepen kunnen worden in het kader van het posttraumatisch stress syndroom. De bevindingen zijn van belang voor de klinische praktijk. Voor een optimale zorg voor de survivors is het belangrijk dat verpleegkundigen, artsen en andere hulpverleners zich bewust zijn van deze symptomen en deze tijdig herkennen. Door het aanbieden van interventie en begeleiding kan men survivors met posttraumatische stress symptomen helpen en hierdoor kan de kwaliteit van leven van deze survivors verbeterd worden.

In *Hoofdstuk 9* worden de belangrijkste bevindingen besproken. Daarnaast worden de implicaties van de onderzoeksresultaten van dit proefschrift geformuleerd voor de klinische praktijk. Daarnaast worden aanbevelingen gedaan voor verder onderzoek. Tot slot wordt de functie van de verpleegkundige op de kinderoncologische follow-up kliniek kort besproken.

Dankbetuiging

In het eerste jaar van het onderzoek las ik de volgende tekst in het boek Death in the afternoon van Ernest Hemingway en ik heb het thuis naast mijn beeldscherm geplakt.

> The great thing is to last and get your work done and see and hear and learn and understand;
> and write when there is something that you know; and not before; and not too damned much after. The thing to do is work and learn to make it.

En dat is wat ik de afgelopen jaren heb gedaan. Ik heb mijn werk zo goed mogelijk proberen te doen. Ik heb opgeschreven wat ik wist en geprobeerd te leren en te begrijpen. Veel mensen hebben mij daarbij terzijde gestaan en dit proefschrift is daarom mede te danken aan de ruime mate waarin zij voor mij beschikbaar zijn geweest. Ik dank de survivors en de personen in de controlegroep die hun medewerking hebben verleend aan het onderzoek. Ik ben hen buitengewoon erkentelijk voor de tijd en de moeite die zij aan de vragenlijst hebben willen besteden. De bereidheid van de huisartsen om te helpen bij het verzamelen van de personen voor de controlegroep wil ik graag afzonderlijk vermelden.

Verder wil ik graag allen bedanken die mij bij dit onderzoek geholpen hebben. De hulp varieerde van het plakken van postzegels op enveloppen, het me op weg helpen als ik weer eens was vastgelopen met mijn analyses tot het tonen van welgemeende belangstelling. Alle vormen waren onmisbaar en ik wil hen hieronder graag persoonlijk bedanken.

Curriculum vitae

Neeltje Elisabeth Langeveld werd geboren in augustus 1954 te Den Helder. Na het voltooien van de middelbare school volgde zij van 1972 tot 1976 de Inservice Opleiding tot A-verpleegkundige in het, toenmalige, Centraal Ziekenhuis in Alkmaar. In dat zelfde jaar verhuisde zij naar Amsterdam waar zij ging werken in het Emma Kinderziekenhuis en in 1977 haar aantekening voor Kinderverpleging behaalde. Hierna werkte zij achtereenvolgens als stafverpleegkundige (1977-1980) en hoofdverpleegkundige (1980-1988) op de afdeling kinderoncologie en volgde zij tussendoor de specialistische opleiding Oncologie Verpleegkunde in het Antoni van Leeuwenhoekhuis en een cursus management (IBW). Na de integratie van het Emma Kinderziekenhuis in het Academisch Medisch Centrum in 1988 werd zij, na een aantal maanden als leidinggevende op de afdeling kinderoncologie gewerkt te hebben, in de gelegenheid gesteld een studiereis te maken naar de Verenigde Staten waar zij diverse ziekenhuizen in Boston, Seattle, San Francisco en Memphis bezocht. Na terugkeer in het Emma Kinderziekenhuis werd zij aangesteld als researchverpleegkundige op de afdeling kinderoncologie. In 1990 begon zij aan de AMC-Opleiding Klinische Epidemiologie voor Verpleegkundigen en deze werd in 1992 afgerond. In februari 1996 werd de Polikliniek Late Effecten Kindertumoren geopend en begon zij aan het onderzoek dat beschreven is in dit proefschrift. Na haar promotie zal zij werkzaam blijven als researchverpleegkundige op de afdeling kinderoncologie. Naast diverse taken binnen het klinisch wetenschappelijk medisch onderzoek en het aansturen van het Klinisch Research Team, richt haar aandacht zich op het opzetten en stimuleren van verpleegkundig patiëntgebonden onderzoek. In januari 2003 is zij daarnaast begonnen met de deeltijd studie Theologie aan de Vrije Universiteit in Amsterdam.

Printed and bound by CPI Group (UK) Ltd, Croydon, CR0 4YY

27/10/2024

14580700-0001